ASIC Design and Synthesis

Vaibhav Taraate

ASIC Design and Synthesis

RTL Design Using Verilog

Springer

Vaibhav Taraate
1 Rupee S T
Pune, Maharashtra, India

ISBN 978-981-33-4641-3   ISBN 978-981-33-4642-0 (eBook)
https://doi.org/10.1007/978-981-33-4642-0

This Springer imprint is published by the registered company Springer Nature Singapore Pte Ltd.
The registered company address is: 152 Beach Road, #21-01/04 Gateway East, Singapore 189721, Singapore

Dedicated to my inspiration and my friend who
wish to create many entrepreneurs in VLSI design

Late Ajit Shelat

Ajit unfortunately passed away on 1st September 2010 in car accident. Due to wish of Ajit I have continued work in ASIC and VLSI design.

Preface

The complexity of the ASIC designs has grown exponentially during the past decade, and during this decade, we are experiencing the AI/ML-based designs and AI-based processor cores to improve the performance of the designs.

The book is the origin of my thought process, and I have tried to document the design concepts and practical issues with their solutions in this book.

The book mainly covers the ASIC design concepts, semi-custom ASIC design flow and the case studies which can be helpful to the postgraduates and professionals. The book uses the Synopsys DC and PT commands and their use during the synthesis and timing closure.

The physical design flow with the basic steps are covered so that readers can have better understanding about the overall ASIC design cycle.

The book is organised in 20 chapters and covers the basics of ASIC design and the concepts used during the RTL design to GDSII flow.

Chapter 1: **Introduction**: Understanding of the ASIC and programmable ASIC plays an important role for the beginners and the experience engineers. The flow can be full-custom, semi-custom, or programmable ASIC, and the major objective of an engineer is to understand the design steps to plan the milestone delivery. The objective of this chapter is to have the basic understanding of ASICs and what should be the focus of the design team.

Chapter 2: **ASIC Design Flow**: The chapter discusses about the ASIC design flow with few of the examples. The chapter is useful to understand about the logic design (frontend design) flow, physical design (backend design) flow and the design flow for the programmable ASICs.

Chapter 3: **Let Us Build Design Foundation**: Understanding of the design abstraction at various levels is always useful during the RTL design and synthesis phase. In this context, the chapter discusses about few of the elements and their use during the design. Even the chapter discusses about the area improvement techniques and role of the design elements in the ASIC design.

Chapter 4: **Sequential Design Concepts**: The objective of this chapter is to have discussion on the synchronous sequential circuits and asynchronous designs and their use during the design phase. For better understanding, the chapter discusses about the sequential elements and their use during design cycle.

Chapter 5: **Important Design Considerations**: The chapter is useful to understand about the basics of timing, skew, latency, and other design considerations such as parallelism and concurrency.

Chapter 6: **Important Considerations for ASIC Designs**: The chapter discusses about these techniques which are useful during the ASIC design architectures and micro-architectures.

Chapter 7: **Multiple Clock Domain Designs**: The chapter discusses about the multiple clock domain designs and strategies which can be useful during the architecture and micro-architecture design.

Chapter 8: **Low Power Design Considerations**: The chapter is useful to understand about the low power design techniques and important strategies which are useful during the ASIC design.

Chapter 9: **Architecture and Micro-architecture Design**: The chapter discusses about the architecture and micro-architecture design concepts and strategies which can be useful during the ASIC design phase.

Chapter 10: **Design Constraints and SDC Commands**: The chapter discusses about the design constraints and the important SDC commands. SDC stands for the Synopsys Design Constraints which is format and used to specify the design intent, including timing, power and area constraints for a design!.

Chapter 11: **Design Synthesis and Optimization Using RTL Tweaks**: The chapter discusses about the ASIC and FPGA synthesis and important concepts useful during the design optimization ad even during RTL design phase.

Chapter 12: **Synthesis and Optimization Techniques**: The chapter is useful to understand the different optimization techniques used during logic synthesis and use of Synopsys DC commands while optimizing the design.

Chapter 13: **Design Optimization and Scenarios**: The chapter discusses about the optimization for the speed and area with the practical design scenarios.

Chapter 14: **Design for Testability**: The DFT and the testability basics for the ASIC design are discussed in this chapter.

Chapter 15: **Timing Analysis**: The STA and the performance improvement techniques are discussed in this chapter.

Chapter 16: **Physical Design**: The chapter discusses about the physical design flow and important issues during the physical design and how to overcome them.

Chapter 17: **Case Study: Processor ASIC Implementation**: The chapter discusses about the overall strategies which are useful during RTL to GDSII for the moderately complex processors. Even the chapter discusses about the performance improvement and the processor architecture strategies with and without pipelining stages.

Chapter 18: **Programmable ASIC**: The FPGA and the role in prototyping is discussed in this chapter. Even the chapter is useful to understand about the FPGA flow and the FPGA synthesis.

Chapter 19: **Prototyping Design**: The design prototyping and the strategies are discussed in this chapter. The use of multi-FPGA architecture, use of the multiple FPGAs during prototyping and the prototyping flow is also discussed in this chapter!

Chapter 20: **Case Study: IP Design and Development**: The IP development and the strategies are discussed in this chapter. The H.264 architecture design and strategies to implement the design is also included in this chapter!

The book is useful to understand the ASIC design flow and the important design concepts useful during the various phases from the architecture design to layout of the chip.

Pune, India

Vaibbhav Taraate

Acknowledgements

The book is originated due to my extensive work in FPGA and ASIC design from year 2000. The journey to develop the algorithms and architectures will continue in future also and will be helpful to many professionals and engineers.

This book is possible due to help of many people. I am thankful to all the participants to whom I taught the subject FPGA and ASIC design, synthesis, and timing closure in few multinational corporations. I am thankful to all those entrepreneurs, design/verification engineers, and managers with whom I worked in the past almost around 20 years.

I am thankful to all my dearest friends and well-wishers for their constant support. Especially, I am thankful to my teammates and all my family members for their support and cooperation!.Thankful to Niraj and Deepesh for their support and cooperation during the manuscript completion phase!

Finally, I am thankful to Springer Nature staff, especially Swati Meherishi, Ashok Kumar, Rini Christy, and Jayarani for their great support during the various phases of the manuscript.

Special thanks in advance to all the readers and engineers for buying, reading, and enjoying this book!

Contents

About the Author

Vaibbhav Taraate is an entrepreneur and mentor at "1 Rupee S T". He holds B.E. (Electronics) degree from Shivaji University, Kolhapur (1995) and received a Gold Medal for standing first in all engineering branches. He completed his M.Tech. (Aerospace Control and Guidance) at the Indian Institute of Technology (IIT) Bombay, India, in 1999. He has over 18 years of experience in semi-custom ASIC and FPGA design, primarily using HDL languages such as Verilog, VHDL and SystemVerilog. He has worked with multinational corporations as a consultant, senior design engineer, and technical manager. His areas of expertise include RTL design using VHDL, RTL design using Verilog, complex FPGA-based design, low power design, synthesis and optimization, static timing analysis, system design using microprocessors, high-speed VLSI designs, and architecture design of complex SOCs.

Chapter 1
Introduction

If we recall the beginning of the miniaturization era, then we can imagine the basic bipolar junction transistor invention at Bell Labs (now AT&T) during year 1947. The first bipolar junction transistor was invented by William Shockley, Bardeen, and Brattain at Bell Labs and got the Nobel Prize in physics during the year 1956. The first integrated circuit was introduced by the 26-year-old engineer Jack Kilby at Texas Instruments (TI).

The popularity of CMOS devices increased during the year 1963 due to low power, high package density, and high-speed requirements.

During the year 1965-1975, Gordon Moore stated Moore's law that is 'Number of transistors in dense integrated circuit doubles in approximate 18–24 months.' The observation of Gordon Moore was valid till the year 2015. It may require modification for the technology node below 10 nm. According to my observation to double the transistors, it may require almost 36 months which may be true during next few decades.

To have basic understanding of the ASIC design and flow, the objective of the remaining sections is to get familiarity with the types of ASICs, different abstraction levels, and few examples which can be useful to think about the ASIC design and strategies.

ASIC is an application specific integrated circuit and designed for specific application.

V. Taraate, *ASIC Design and Synthesis*,
https://doi.org/10.1007/978-981-33-4642-0_1

1.1 ASIC Design

The era of miniaturization from the year 1960 to 2020 has witnessed lot of the evolutions and design changes. What we need to understand is that, what is exactly ASIC design? Now consider the small square of few micrometers or nanometer which is empty box. Now for the specific functionality, design team will fill this empty box with functional blocks. The design team which performs this is front-end (logic) design team.

The backend or physical design team works in the area of floor planning to physical verification at the chip level for the specific technology node.

The manufacturing unit which is foundry performs the manufacturing and packaging of the chip in mass, and initially few sample pieces will be tested by the design houses to understand the intended design outcome.

Now how all above is achieved? All the design-related work is achieved by the intelligent chip designers working in the various areas of the chip design using the Electronic Design and Automation (EDA) tools. The various popular EDA tools are from Synopsys and Cadence and extensively used in the design of chip and to improve or to achieve the desired performance.

With the functionality of the chip, it requires to understand about the constraints such as area, speed, and power and the main goal of the logic design and physical design team is to understand the block-level and top-level constraints and to have the better strategy to achieve the desired performance.

For the basic understanding, consider the pipelined processor which performs the basic arithmetic and logic operations such as addition, subtraction, multiplication, division, XOR, OR, AND, and NOT. What we need to imagine at the higher level is the functional blocks, complexity of the design that is rough estimation of the area, what constraints we should apply, and what exactly we will achieve. At the beginning, we will have just the basic idea of the blocks and as the design evolves we will land up into the phase of the chip architect.

For the idea of the architecture of the above chip, the basic layout is shown in Fig. 1.1. The subsequent chapter discusses the design flow, chip architecture, micro-architecture!

1.2 Types of ASIC

ASIC Design: ASIC stands for the application-specific integrated circuit and is designed to perform the specific application. For example, the processor or controller is used to process the specific information.

The following are different types of ASICs

1. Full-custom ASIC
2. Semi-custom ASIC
3. Gate array-based designs
4. Structured ASICs.

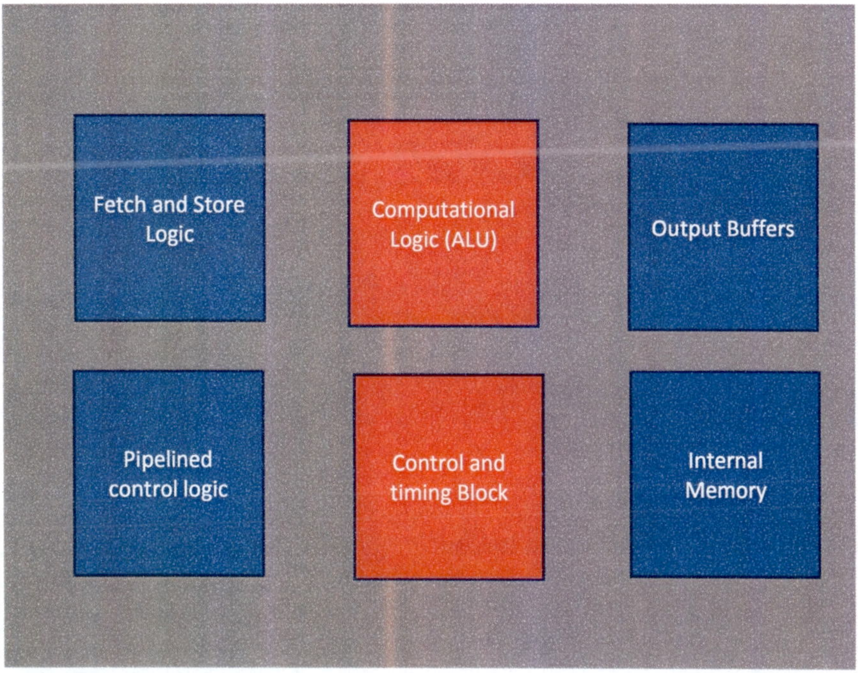

Fig. 1.1 Basic chip layout

Full-Custom ASIC: In this type of ASICs, the design starts from scratch for the specific technology node. Each cell is designed depending on the technology node requirements. This approach is useful for high volume production, and one can imagine the microprocessors and floating-point processors, which are required in the design and can be designed using the full-custom design flow.

The major advantage of the full-custom design is that for the high volume production it gives the lower power, high speed, and the least gate count. Achieving the constraints of the speed, area, and power is time consuming for this flow. But as the cells are designed from the scratch for the desired technology nodes, the desired constraints can be achieved.

The major disadvantage of this flow is the high non-recurring expenditure and the long design cycle time.

Standard Cell-Based ASICs

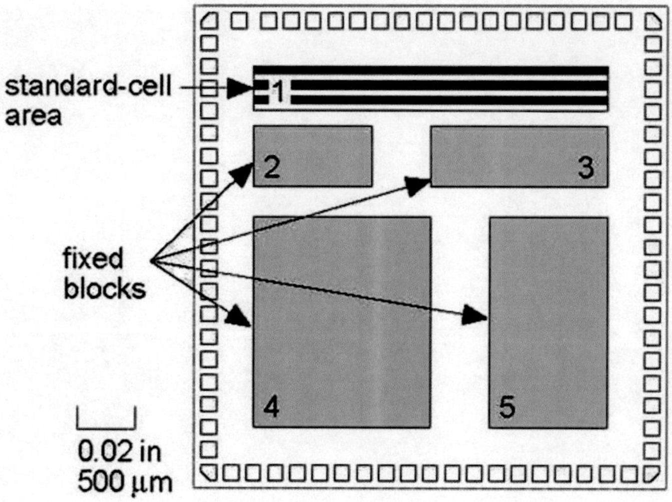

In this type of design flow, the standard library cells such as NAND, NOR, XOR, and flip-flops are used during the design. The beauty of this flow is that it uses the pre-defined and prefabricated cells, for example, RAM hard macro-cores, etc. The transistors and interconnects are customized; that is, all the mask layers are customized.

The advantage of this flow is that, as compared to full-custom ASIC the design cycle time is shorter and for the specific technology node the pre-validated standard cells like microprocessors and macros are available during the design.

The disadvantage is that as compare to the gate array based ASICs, the design has the high NRE and it needs separate fabrication mask for each design.

Gate Array-Based ASIC

In this type of ASICs, the wafers are prefabricated with unconnected gate array. That is, wafers are common for all the designs. The types of gate array-based ASICs are mainly of following two types.

1. Channeled gate array
2. Channel-less gate array.

Channeled Gate Array: In this type of ASIC, the interconnects use the pre-defined spaces between the rows of base cells.

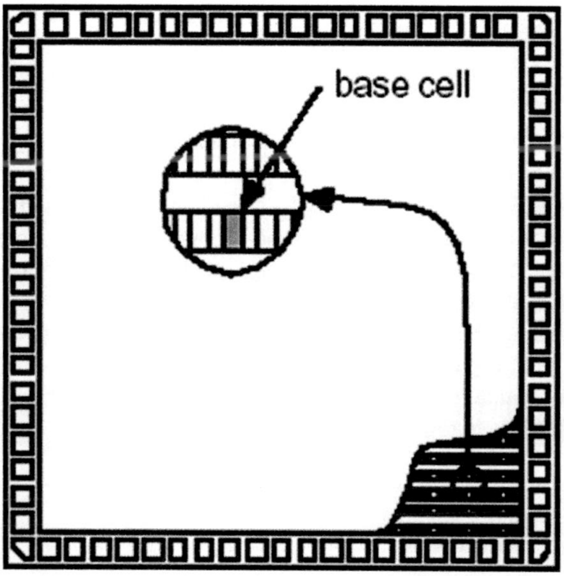

Channel-Less Gate Array: In this type of ASICs, the few top mask layers are customized.

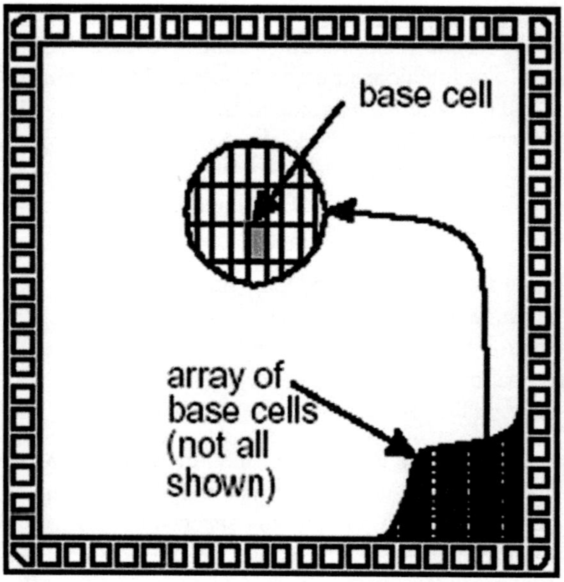

The major advantage of the gate array-based ASIC is that, lower NRE cost as the same wafer is fabricated for the multiple designs. Another main advantage is the low turnaround time.

The main disadvantages are the low density, lower volume, and the less optimized design.

Structured ASICs

A structured ASIC falls between and gate array and a standard cell-based ASIC.

The main design task involves mapping the design into a library of building block cells and interconnecting them as necessary. The main important points regarding the structured ASICs are components are 'almost' connected in a variety of pre-defined configurations and only a few metal layers are needed for fabrication which in turn drastically reduces turnaround time.

The advantages of the structured ASIC are low NRE cost, less complexity, low power consumption, high performance, and the smaller marketing time.

The main disadvantage is that the team needs to have better understanding of the design constraints due to the use of prefabricated design cells.

1.3 Abstraction Levels

The design can go through the different abstraction levels such as functional design, logic design, gate-level design, and the switch-level design. This section discusses these abstraction levels in more detail.

1. **Functional Design**: Now imagine a scenario to design a product or chip, so the first thought is the product idea or/and depending on the idea the chip functionality can be extracted. The functional design is basically the outcome of the functional specification, and the group of team members can create the high-level and low-level design document and can code the functionality using the higher-level language such as C or C++. For example, consider the H.264 encoder design, and the functional design team can create the golden reference model using the high-level language by using the following

   (a) The types of frames which need to be processed
   (b) The frame support
   (c) The prediction blocks and functionality
   (d) The quantization and transform algorithms required
   (e) The entropy coding methods

   If the desired functionality is validated, then the design can be considered as golden reference model which can be used throughout the design.
2. **Logic Design**: The logic design team understands the architecture of the chip and the partitioning mechanisms to complete the RTL design, where RTL stands for the register transfer level. The team of professionals uses the HDL such as VHDL, Verilog, and SystemVerilog to have the RTL design and verification at the block and top level. The main advantages of the HDL used during the RTL design are as follows:

(a) The HDL supports the concurrent and sequential constructs.
(b) HDL supports the notion of time.
(c) HDL supports describing the interfaces and ports as input, output, bidirectional.
(d) HDL supports the edge- and level-sensitive design constructs.

For more details about the RTL design and verification, refer Chaps. 3 and 4. The logic design flow is discussed in Chap. 2.

The RTL design example using the Verilog to infer the 2-bit shift register using the non-blocking assignment is described in Example 1.

Example 1 The RTL description using Verilog

```
/////////////////////////////////////////////////////////////////////////////
module non_blocking_assignments
 ( input data_in, clk, reset_n,
  output reg data_out);

reg tmp;

always @ ( posedge clk or negedge reset_n)
begin

        if ( ~ reset_n)
        begin
                { data_out, tmp } <= 2'b00;

        end

        else
   begin

            data_out <= tmp;

                     tmp <= data_in;
                end

end
endmodule

/////////////////////////////////////////////////////////////////////////////
```

3. **Gate-Level Design**: The RTL is given as one of the inputs to synthesis tool to get the gate-level netlist. The synthesis is process of getting the lower level of abstraction from the higher-level design (Fig. 1.2).

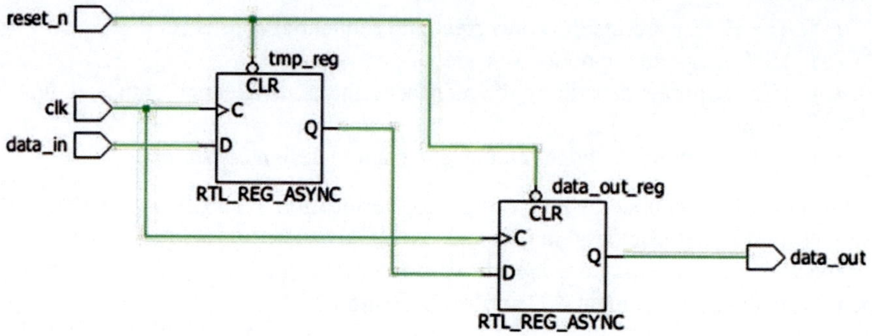

Fig. 1.2 RTL schematic of Example 1

4. **Switch-Level Design**: The design using the CMOS standard cells and switches is called as switch-level design. In the simple term, the physical design or backend design is like playing with the switches and standard cells, macros for the specific technology node. The backend/physical design flow is discussed in Chap. 2.

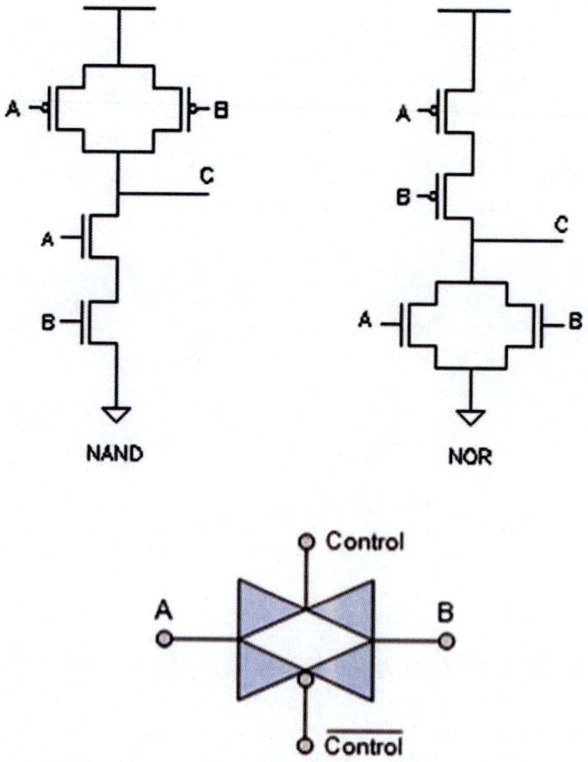

1.4 Design Examples

Now consider the ASIC design of H.264 encoder and decoder, and what we should do?

1. Market survey to understand the availability of various products in the market
2. The functional specification of the H.264 encoder and decoder
3. The functional design documentation such as high-level design (HLD) and low-level design (LLD) and design planning
4. Logic Design: Plan the design
   (a) Specification understanding and the architecture design
   (b) RTL design and verification
   (c) Synthesis/DFT and timing verification
5. Physical Design: Design from floor planning to physical verification
   (a) Planning of the design (floor planning and power planning)
   (b) CTS
   (c) Place and route
   (d) Physical and timing verification
   (e) GDSII
6. Manufacturing and Test: The design manufacturing and test phases
   (a) Fabrication
   (b) Packaging
   (c) Test.

Consider the initial floor plan of the H.264 encoder (Fig. 1.3).

1.5 What We Should Know?

During the ASIC design cycle with functional design and validation, we should focus on the area, speed, and power constraints.

1. **Area**: The chip area and the logic area which defines the overall density of the design in the few micrometer square. Meeting the area constraints is one of the important tasks during the logic and physical synthesis. The area optimization can be achieved at various levels such as
   (a) Architecture tweaks
   (b) RTL tweaks
   (c) Using synthesis commands
   (d) At the physical design using dedicated cells

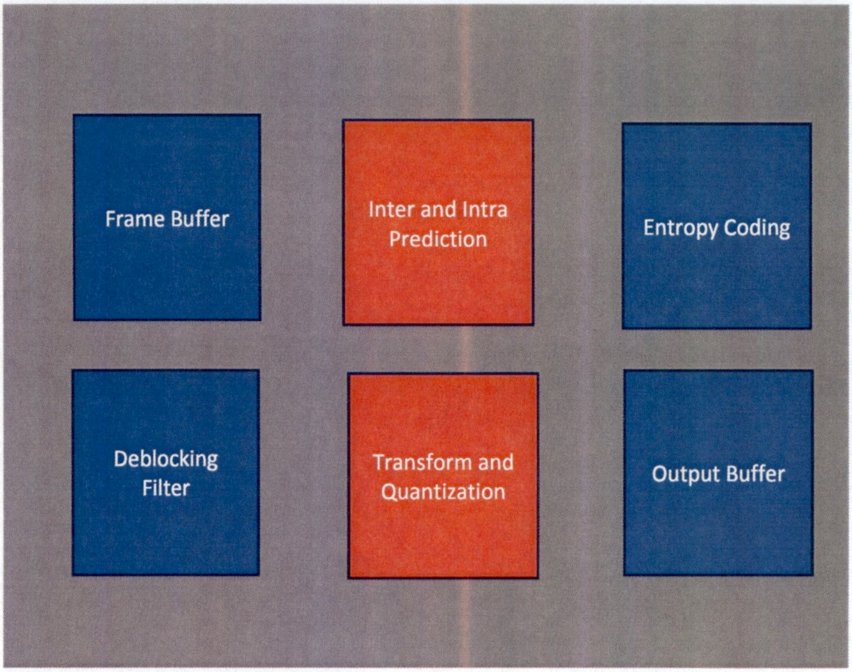

Fig. 1.3 Initial floor plan of H.264 encoder

2. **Speed**: The speed is another important constraint and can be achieved at the block and top level using the
   (a) Synopsys PT commands
   (b) RTL tweaks
   (c) Architecture tweaks
   (d) During physical design
   (e) Using the dedicated IPs
3. **Power**: The power, static and dynamic are another important constraints for ASIC and can be achieved using the following
   (a) Use of low-power architecture
   (b) Low power cells
   (c) RTL tweaks to reduce the dynamic power
   (d) Low power format
4. **Clock Skew**: The skew is the difference between the clock arrivals at two different coordinates.
   (a) **Positive Clock Skew**: The launch flip-flop is triggered first and then capture flip-flop.

   (b) **Negative Clock Skew**: The launch flip-flop is triggered last, and capture flip-flop is triggered first.

5. **Slack**: The slack is the difference between the two different time instances.
   (a) **Setup Slack**: The setup slack is the difference between the data required time and data arrival time.
   (b) **Hold Slack**: The hold slack is the difference between the data arrival time and data required time.
6. **Clock Gating**: The use of the clock gating cells to minimize the dynamic power.
7. **Synchronous Design**: All the flip-flops in the design are triggered by using the common clock source.
8. **Asynchronous Design**: The flip-flops in design are triggered by the different clock sources.

1.6 Important Terms Used Throughout Design Cycle

The following are few important terms which we should know during the ASIC design cycle.

1. **Architecture**: Block-level representation of design
2. **Micro-architecture**: Sub-block-level representation of design
3. **RTL**: Register transfer level
4. **RTL design**: Design using the HDL synthesizable constructs
5. **RTL verification**: The testbench and automation using the non-synthesizable constructs
6. **Synthesis**: The process of getting the gate-level netlist from the RTL. Or it is the process of getting the lower level of abstraction from the higher-level design
7. **DFT**: Design for test to find the manufacturing defects
8. **STA**: Static timing analysis at the pre-layout or post-layout
9. **Floor planning**: The chip floor plan
10. **Power planning**: The power mesh and ring planning for the chip
11. **CTS**: Clock tree synthesis, the clock trees for uniform distribution of skew, and the strategy
12. **P and R**: Placement and routing that is the placement of standard cells and macros, IPs, and to route them
13. **Physical verification**: The verification that is LVS and DRC
14. **LVS**: Layout versus schematic check
15. **DRC**: Design rule check
16. **Back annotation**: The RC extraction
17. **GDSII**: GDSII stream format, common acronym GDSII, is a database file format which is the de facto industry standard for data exchange of integrated circuit or IC layout artwork.

1.7 Chapter Summary

The following are few important points to conclude the chapter.

1. 'Number of transistors in dense integrated circuit doubles in approximate 18–24 months' is called as Moore's law.
2. ASIC stands for the application-specific integrated circuit.
3. FPGA stands for the field programmable gate array.
4. The major advantage of the full-custom design is that for the high volume production it gives the lower power, high speed, and the least gate count.
5. The main advantage of the semi-custom ASIC is, as compared to full-custom ASIC the design cycle time is shorter and for the specific technology node the pre-validated standard cells like microprocessors and macros are available during the design.
6. The major advantage of the gate array-based ASIC is that lower NRE cost as the same wafer is fabricated for the multiple designs.
7. The main design constraints are area, speed, and power.

Chapter 2
ASIC Design Flow

As discussed in the previous chapter, the ASICs can be of type full-custom, semi-custom, gate array-based ASICs. The major objective of the following few sections is to have the detailed discussion about the semi-custom ASIC design flow and the programmable ASIC design flow. The important design examples are also discussed in the next few sections and useful during the ASIC and FPGA design.

2.1 ASIC Design Flow

The semi-custom ASIC design in which the standard cells and macros which are pre-validated is used. As discussed in Chap. 1, we can have different types of ASICs such as full-custom, semi-custom, gate array-based and depending on the design requirements we can choose one of the flows. Figure 2.1 describes few of the important design phases during the ASIC design cycle.

1. **Market Survey and Specification Extraction**: It is one of the important phases during the design cycle. Before the logic design, the team performs the market survey to understand what are the different products of similar type available in the market. Origin of any design idea or product can be realized in the quick time, and product should be excellent in all the aspects that are the major objective of any organization. The excellence in the design and product innovation is objective of many research and development organizations. For example, consider Intel as processor design organization, what they work on the processing capability, low power architecture design, high-speed designs, signal integrity, and more reliability to their chipsets.

Understanding of the ASIC design flow plays an important role during design cycle.

V. Taraate, *ASIC Design and Synthesis*,
https://doi.org/10.1007/978-981-33-4642-0_2

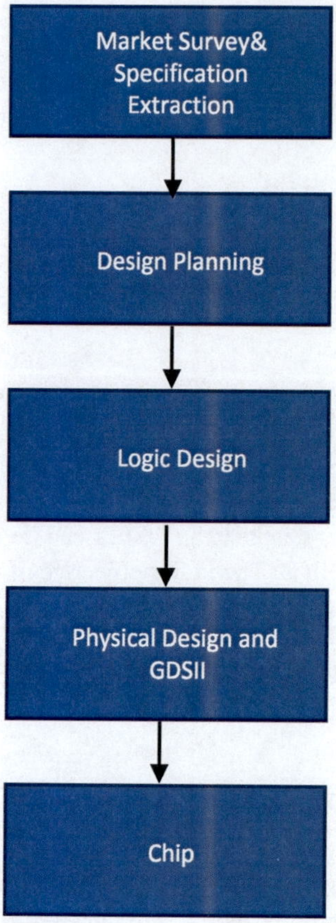

Fig. 2.1 Semi-custom ASIC design flow

For new idea, finalization of the specification and the architecture of the chip is the primary task and for that the market survey plays an important role. Following the team does during the market survey.

(a) The detailed understanding of the available products
  (i) Understanding of functionality, speed, power, and area
  (ii) Understanding about the electrical characteristics
  (iii) Understanding about the mechanical assembly and packaging
  (iv) Understanding about the user interfaces
(b) The volume and cost of product
(c) End customer base
(d) How the new idea can be better as compared to existing products?

The main outcome of all above is to extract the specifications of the product or chip at various levels. Our goal is to work on the functional design of the chip so we will consider the functional specifications.

Let us consider the 32-bit processor what we need is the following!

(a) The operations performed by the processor such as arithmetic, logical, data transfer, branching, and floating point
(b) The complexity of bus interfaces such as address bus and data bus
(c) The performance improvement mechanism such as pipelining and the configuration support
(d) The electrical characteristics of interfaces such as slew, voltage levels, and power
(e) The external interface information and compatibility
(f) The internal storage information and the data computation schemes
(g) The IP availability and their specifications
(h) The technology node for ASIC and the performance
(i) What are the constraints achieved such as area, speed, and power?

By using all above, the specifications of the product can be documented and the team understands the feasibility of the product using few risk and dependability matrix parameters. Consider the parameter as a speed, existing chipset is working on the operating frequency of 400 MHz and technology node is 10 nm then can the desired product can operate at 450 MHz or not?
If answer is no due to the technology library cell characteristics, then the choice can be work on the 400 MHz but with more parallel computational elements or can use the lower technology node such as 7 nm to achieve the 450 MHz operating frequency.

2. **Design Planning**: The design planning in technical terminology is the architecture and micro-architecture design, but practically with this we need to work on the project planning and chip delivery plan to the market. So broadly concurrent teams of technical, man management, and the delivery will work to accomplish the design task. The project planning is ruled out as per as discussion in this book is concerned. Our objective is to work on the understanding of the specification to have the top-level architecture in place so that we can plan for the logical, physical design and chip manufacturing and test phases.
The specification extraction document is used as input during this phase to get

(a) Architecture design and micro-architecture design
(b) Architecture tweaks to estimate the rough area and the possibility of achieving constraints
(c) Having information about the top-level interfaces and the timing
(d) Can be used to understand about the storage and memory requirements
(e) Useful for the project planning and planning of milestone delivery

We will consider the outcome of the design planning stage as the architecture and micro-architecture evolution to have the better architecture design.

3. **Logic Design**: The logic design phase of the ASIC is very important as the quality of the RTL design and verification decides about the quality of chip. During the logic design phase, broadly we need to perform the following and discussed in Sect. 2.1.1.

   (a) RTL design
   (b) RTL verification
   (c) Synthesis
   (d) DFT and scan insertion
   (e) Equivalence checking
   (f) Pre-layout STA

4. **Physical Design and GDSII**: The physical design phase of the ASIC is also called as backend design, and the physical design team uses the gate-level netlist as one of the inputs with the technology libraries to get the GDSII. During the physical design phase, broadly we need to perform the following and discussed in Sect. 2.1.1.

   (a) Floor planning
   (b) Power planning
   (c) CTS
   (d) Place and route
   (e) LVS
   (f) DRC
   (g) Signoff STA
   (h) GDSII

5. **Chip**: To get the chip from foundry, there are several manufacturing and packaging processes. The sample chips will be issued to test houses to perform the testing.

2.1.1 Logic Design

The logic design flow uses the functional design specifications and the target technology library for the ASIC to get the gate-level netlist. The logic design flow is shown in Fig. 2.2 and consists of the important steps as follows

1. Design partitioning at the architecture level
2. RTL design and verification
3. Synthesis and DFT
4. Equivalence checking
5. Pre-layout STA.

This section discusses these important milestones.

1. **Design Partitioning**: As the ASIC design architecture is complex in nature and consists of million or billion gates, the better idea is to have the partitioning of the design at the architecture level. The architecture and micro-architecture document is evolved from the functional specifications and is used as reference for the design.

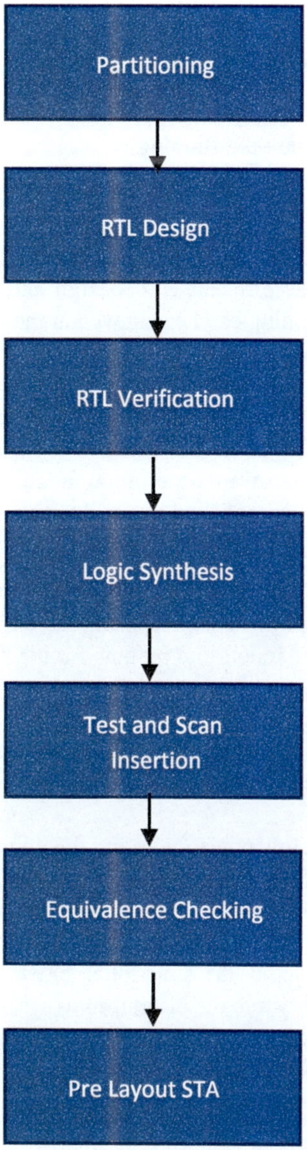

Fig. 2.2 ASIC logic design flow

The design partitioning goal is to have the modular design approach to get the better quality of the RTL. The design partitioning team considers the following points while partitioning of the design.

   (a) The complexity of the blocks
   (b) IPs used
   (c) Single clock verses multiple clock
   (d) Low-power design strategies
   (e) Software and hardware partitioning
   (f) Constraints for the functional blocks

   By considering all above, the design is partitioned into multiple blocks and used as the basic foundation for the RTL design and verification. Better design partitioning can yield into the efficient RTL design and in turn the chip development. The design partitioning can be achieved at the different levels such as the architecture, RTL, and netlist level.

2. **RTL Design**: As discussed in Chap. 1, the RTL is register transfer level and it is the representation of the design functionality using the hardware description language. Throughout this book, our focus is to have discussion on the ASIC design using Verilog as hardware description language. During the RTL design phase, the following is accomplished by the design team.

   (a) Understanding of the functionality of the design and the design partitioning
   (b) The block-level RTL design. If the block functionality is complex, then the micro-architecture of the design is to implement the sub-block RTL
   (c) Use of the registered inputs and outputs for the better and clean timing
   (d) Use of the area and speed improvement features at the RTL level. The features such as resource sharing and pipelining
   (e) The RTL design using the modular design approach
   (f) Use of the synchronizers for passing of the data between the different clock domains
   (g) Use of the required low power cells and the power formats during the RTL design stage
   (h) Understanding of the use of required IPs and the integration to get the desired functional output
   (i) Top-level integration of the functional block to yield into the desired intended functionality and the clean timing
   (j) Sanity-level verification to confirm the design functionality and functional correctness of the design

3. **RTL Verification**: For any kind of chip, the functional correctness at the block, top, and chip level should be achieved and that is the primary objective of the verification team. The following are few of the important tasks which verification team should complete during the verification cycle.

   (a) Verification planning for the chip
   (b) Verification architecture
   (c) Creating the test cases and test vectors which can be used during the verification of the block and top functionality
   (d) Assertion-based verification mechanism and automation in the testbenches to achieve the desired coverage goals

   (e) Efficient reporting mechanism and better communication strategies with the RTL design team
   So in the simple way what the verification team does is that better verification plans and strategy to check for the functional correctness of the block and tip-level design using the robust and automated testbenches!

4. **Logic Synthesis**: The synthesis is the processes of getting the lower level of abstraction from the higher level. During the logic design, the objective is to get the gate-level netlist from the Verilog RTL. During this book, our goal was to have the focus on the FPGA synthesis using the Xilinx EDA tool Vivado, and ASIC synthesis using the Synopsys Design Compiler which is popular EDA tool and mostly referred as Synopsys DC.
   The ASIC synthesis tool uses the following inputs to get the gate-level netlist.

   (a) RTL design (.v files)
   (b) Optimization constraints such as area, speed, and power
   (c) Technology library

   The synthesis tool should be efficient enough to achieve the target constraints such as area, speed, and power. If constraints are not achieved, then it is recommended to have the architecture tweaks and RTL tweaks.
   Few of the architecture tweaks can be

   (a) Improve the partitioning strategies
   (b) Use the pipelined design approach
   (c) Partition design at the sequential boundaries
   (d) Include the parallelism by replicating design into multiple blocks

   Few of the RTL design tweaks can be

   (a) Use resource sharing
   (b) Use dead zone elimination and constant folding
   (c) Use the pipelining for the clean timing
   (d) Use the sequential boundaries and modular approach during the block-level design

   The netlist can be saved in the file format and can be .v file or the Synopsys database.
5. **Test and Scan Insertions**: The design for testability is used to get the manufacturing defects from the design. The DFT is mainly to find the stuck at faults from the design may be single, double, or triple stuck at faults. The DFT is mainly to check for the controllability and observability of the different nodes in the design. The DFT is an important milestone to get the information about the defects. The DFT schemes can be of various types, and few of them are

   (a) Ad hoc DFT
   (b) Structured DFT

   (i) Scan-based DFT
      1. Partial scan
      2. Full scan
   (ii) Built in Self-test
      1. LBIST
      2. MBIST
   (iii) JTAG

   The DFT schemes and the case study are discussed in the subsequent chapters, and the major goal is to understand about the DFT schemes with the objective and the role of EDA tools.
6. **Equivalence Checking**: To preserve the design intent, the equivalence checking needs to be performed to check the logic equivalence. The equivalence checking uses the formal verification techniques.
   The objective of the equivalence checking is to verify the RTL design functionality.
7. **Pre-layout STA:** The STA stands for the static timing analysis and is one of the important milestones during the logic design. The major goal is to find out the timing violations from the design, and these violations are mainly the setup and hold time. During the pre-layout STA, the setup time violations are fixed as the design doesn't have the routing information. STA popular EDA tool is Synopsys PT (PT shell) and used to report the timing information by analyzing all timing paths from the design. The STA tool uses the input as the gate-level netlist, timing library, and technology library with the timing constraints to analyze the timing for the design.

More about the timing is discussed in the subsequent chapters.

2.1.2 Physical Design

The physical design flow with the Synopsys IC Compiler is discussed in Chap. 16. This section discusses what exactly the design team works during the physical design phase.

As the gate-level netlist is available from the logic design flow, the netlist with chip constraints and required libraries is used as input during the physical design flow.

The physical design starts with the floor planning that is planning of the design mapping, and the goal is that there should not be congestion while routing of the design and the logic blocks or functional blocks should meet the aspect ratio. The better floor planning is required to have the better area, speed, and power and will be useful to avoid the routing congestion. The power planning stage is used to plan for the power rings (VDD and VSS) and power straps depending on the power requirements (Fig. 2.3).

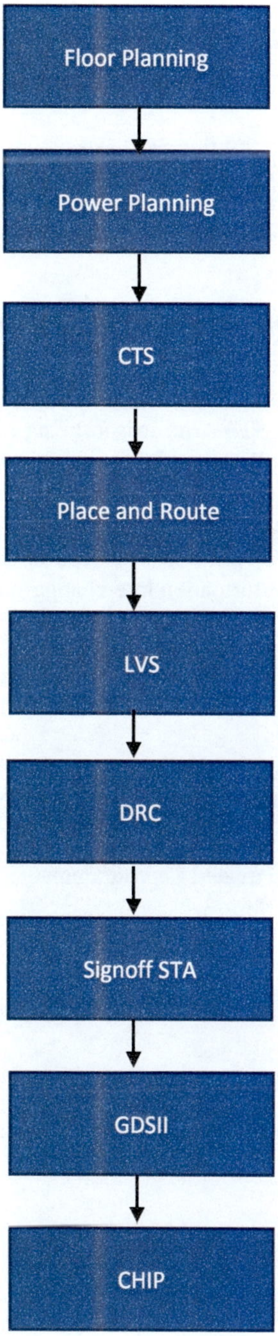

Fig. 2.3 ASIC physical design flow

After the power planning is done, the clock tree synthesis needs to be performed to balance the clock skew and to distribute the clock to the various functional blocks of the design. The clock tree can be H tree, X tree, balanced tree and discussed in Chap. 16.

The placement and routing are done to have the layout of the chip. The layout will have the routing delays, and many times the STA needs to be performed to find and fix the timing violations.

The layout of chip needs to be checked to verify the

(a) Foundry rules, that is, DRC
(b) LVS that is checking of the layout versus the schematic and the intent is to verify the layout with the gate level netlist.

If all the design rules are met and there are no any issues in the LVS, the team needs to perform the signoff STA. The reason being after the layout, most of the time the design will not meet the required timing and frequency and may requires the modification or tweaking at the various stages. The flow is iterative, and objective is to achieve the chip-level constraints.

After the signoff STA the GDSII is generated. The GDSII stands for the Generic or Geometric Data Structure Information Interchange and describes the layout of the design with the connectivity.

The foundry uses the GDSII to manufacture the chip. It is also treated as tapeout delivered to foundry!

2.2 FPGA Design Flow

FPGA design flow can be also treated as programmable ASIC flow and described in Fig. 2.4.

The important steps are

1. Design planning
2. RTL design and verification
3. Synthesis
4. Design implementation: It consists of the following steps
    (a) Logic functionality mapping
    (b) Place and route
    (c) SDF-based verification
    (d) Signoff STA
5. Device programming.

Few of the important FPGA blocks are shown in Fig. 2.5. The modern FPGA architecture is complex and consists of few of the important blocks:

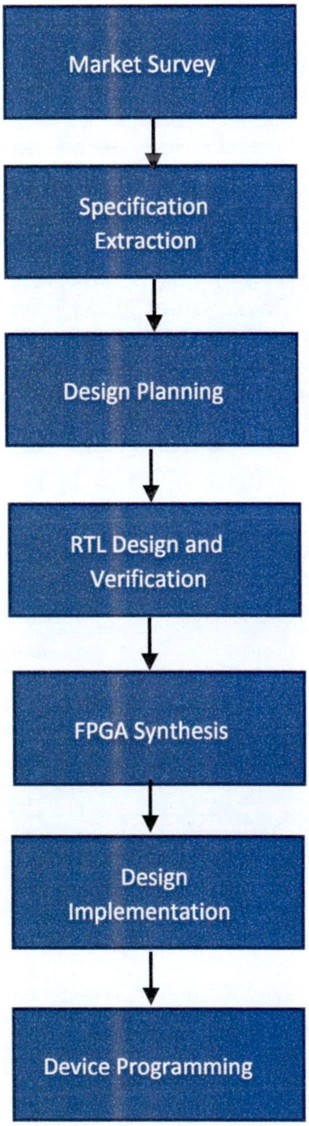

Fig. 2.4 FPGA design flow

1. Configurable logic block (CLB)
2. IO blocks
3. Switch boxes
4. DSP blocks
5. Multipliers
6. Processor block.

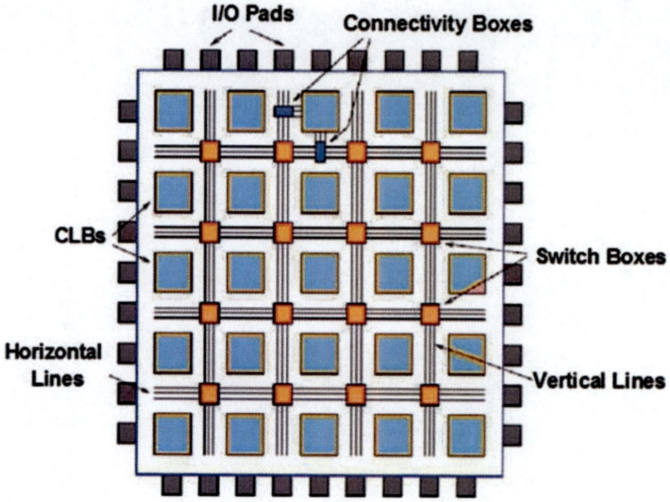

Fig. 2.5 FPGA architecture

Table 2.1 Comparison between ASIC and FPGA

Important parameters	ASIC	FPGA
Cost per piece	Low	High
Time to market	Slow	Fast
NRE cost	High	Low
Size	Small	Medium
Design complexity	Very complex	Moderately complex
Power consumption	Low	High
Performance	High	Moderate

The comparison between the ASIC and FPGA is shown in Table 2.1 by considering few important parameters.

2.3 Examples and Thought Process

Before the ASIC or FPGA design what we need to work on or what we need to revise as foundation is discussed in this section. Consider the complex ASIC what we should understand.

1. The role of ASIC and purpose of design
2. What is technology node
3. Functional and timing details

4. Important design elements such as standard cells, micros, IPs for the ASIC-based designs. Functional blocks and architecture for FPGA-based designs
5. What are the constraints ?
6. For semi-custom ASIC, the reusable elements and components
7. We should think about the
   (a) Area, speed, power requirements
   (b) Data bandwidth
   (c) Latency and clock requirements
   (d) Multiple clocks in design and synchronizer requirements
   (e) Low-power architecture and sequencing
8. Complexity of RTL design, verification, and available IPs
9. EDA tools and optimization efforts required
10. Testing requirements and test planning.

2.4 Design Challenges

If we consider the complex ASIC with million gate count, then important challenges are

1. Design partitioning
2. Longer time to market
3. Complexity in the design and verification
4. Large number of resources
5. Availability of the functional and timing proven IPs
6. Meeting the optimization and design constraints
7. Interconnect delays and requirement of high-speed IOs
8. Testing time and the requirements
9. Effect of noise and the field testing
10. OCV analysis and chip tests.

The subsequent chapters discuss the practical issues in the ASIC design and how to fix them at various stages of the ASIC design cycle.

2.5 Chapter Summary

The following are few of the important points to conclude the chapter.

1. Important logic design flow steps are: RTL design, RTL verification, synthesis, DFT and scan insertion, equivalence checking, and pre-layout STA.
2. Important physical design flow steps are: floor planning, power planning, CTS, place and route, DRC, LVS, signoff STA, GDSII.

3. During the logic design phase, the information about the clock network is not available.
4. The layout of ASIC is floor plan, placement, and routing of the design.
5. During the pre-layout STA, the objective is to fix the setup violations.
6. After the P and R, the overall timing and frequency for the design will never meet and hence post-layout STA is required.
7. Layout of the ASIC chip is the deliverable as the timing is clean and desired constraints are met!.
8. The DRC is performed to confirm whether all foundry rules are met or not!
9. GDSII is Generic Data Structure Information Interchange and delivered to foundry to manufacture ASIC.
10. The RTL to GDSII flow is basically ASIC design flow which consists of the logic design and physical design.

Chapter 3
Let Us Build Design Foundation

During the RTL design phase, we always use the synthesizable constructs, and during the synthesis phase, we will try with objective to achieve the required constraints. Synthesis is a process to get the lower level of abstraction from the higher level of design. During the ASIC design, we will get the various design abstraction levels such as

1. Functional design
2. Architecture and logic design
3. Gate-level design
4. Switch-level design
5. Physical design.

To understand these various abstraction level, the chapter discusses about the important design elements used frequently during the logic design.

3.1 Combinational Design Elements

In the combinational design, an output is function of the present input only. That is if input changes, then an output will change. In the practical scenario, the output will not change instant immediately, but it will change after some delay which we can treat as the gate delay or the propagation delay.

Understanding of the basic design elements plays important role during the logic design and synthesis.

V. Taraate, *ASIC Design and Synthesis*,
https://doi.org/10.1007/978-981-33-4642-0_3

Table 3.1 Combinational design elements

Combinational logic elements	Description
Universal logic gates	NAND, NOR universal logic gates to implement the combinational design or Boolean function
XOR, XNOR	Used to perform the complement of input and used in parity detectors
Arithmetic resources	The adders, subtractor, multipliers used as the arithmetic resources
Multiplexers	Multiplexers have many inputs and single output and used as data selector or as universal logic to implement the Boolean function. Useful during the pin multiplexing and clock multiplexing
Demultiplexers	Demultiplexers have single input and many outputs and used as de-multiplexing logic. Useful during the pin de-multiplexing and clock de-multiplexing
Decoders	The decoders are extensively used to decode the data and to enable one of the functional block at a time
Encoders	Encoder is used to encode the binary data and performs the reverse operation as compare to decoder
Priority detectors	Priority detectors or priority encoders we can use to assign the output depending on the priority of the level sensitive inputs

The propagation delay of the logic gate is the amount of time required for the logic gate to generate the valid output when input changes either from 1 to 0 or vice versa. The propagation delay is defined for the standard cell or logic gate in the timing library.

The important elements are designed using the logic gates, and we mainly consider the NAND and NOR logic gates as universal gates. The important combinational elements are multiplexers, demultiplexers, decoders, encoders, code converters, and the arithmetic logic elements adders, subtractors, and multipliers.

Table 3.1 discusses about the important combinational elements and their role during the design phase and useful to design the glue logic, functionality for the specific requirement. For few of these elements, the design strategy is explained in this chapter and useful during the RTL design phase.

The following section discusses about these logic elements in the context of the ASIC and FPGA design.

3.2 Logic Understanding and Use of Construct

Understanding the functionality for the individual functional block is especially important during the ASIC design cycle. Whether it is moderate gate count design or the complex design, the understanding will play important role to choose the suitable construct during the RTL design phase. For combinational modeling, we will choose the 'assign' construct. The 'assign' is continuous assignment, and it is neither blocking nor non-blocking. The multiple assign constructs will infer the combinational logic, and the multiple continuous assignments will execute concurrently.

The Syntax is

```
assign expression_1;
assign expression_2;
```

where ***assign*** is keyword. The keywords used in the Verilog (.v) file are marked using bold characters.

For example, consider the following continuous assignments

```
assign sum = a_in ^ b_in;
assign carry = a_in & b_in;
```

Both above continuous assignments will execute and update in the active event queue, and they will execute concurrently to infer combinational logic as XOR and AND gate.

3.3 Arithmetic Resources and Area

Let us try to discuss the arithmetic resource which is the adder and subtractor. To optimize the area, the subtraction operation is implemented using the adders that is by using the 2's complement addition. For example, if we have 8-bit binary input A then A − 1, we can consider as A + 11111110 + 1=A + 1111_1111.

Now, let us discuss the reason for performing the subtraction using the 2's complement addition! Consider a practical scenario where we need to perform the 4-bit addition and subtraction where A is 4-bit binary input, B is 4-bit binary input and Control_Input controls the operation. Table 3.2 indicates these two operations.

Table 3.2 Description of the logic

Control_Input	Operation	Description
0	Addition (A, B)	A + B
1	Subtraction (A, B)	A − B

Area Requirement: If we don't use the 2's complement for the subtraction, then we need to use the resource as adder and subtractor and some data selection logic to generate the output. So more the resources, you can imagine the logic inferred as shown below. As shown, the design needs two 2:1 MUX, 4-bit adder and 4-bit subtractor.

Issues in the Design: The main issue in the design strategy is poor data path management and more resources in the data path. Both the adder and subtractor are performing the operation at a time, and depending on the Control_Input the output Result, carry/borrow is updated. So, the poor data path optimization due to lack of sharing the common resources. Another issue is the use of multiplexing logic at the outputs (Fig. 3.1).

Let us optimize the above logic using the concept of sharing of common resources. Now, to eliminate the data selection logic that is 2:1, multiplexer and subtractor let us use the full adder as arithmetic resource. The subtraction can be performed using the 2's complement method, and to perform A − B, use the A + (~B) + 1 that is perform 1's complement of B using the XOR logic. If one of the inputs of XOR gate is logic '1', then output of XOR is complement of the present input (Table 3.3).

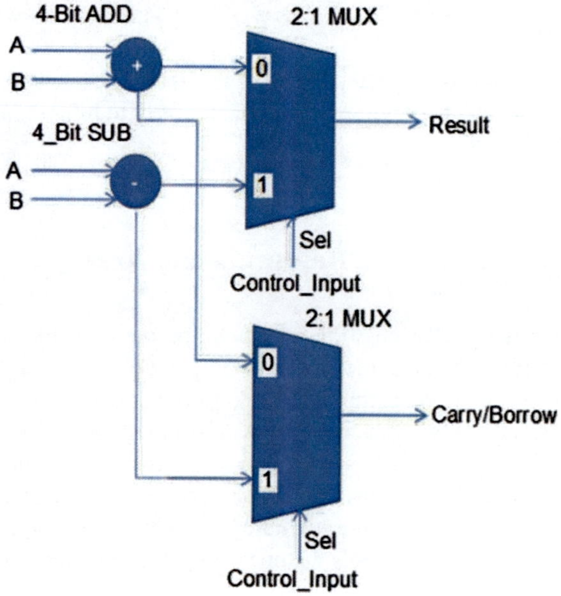

Fig. 3.1 Logic without resource sharing

Table 3.3 Description with goal of resource sharing

Control_Input	Operation	Description
0	Addition (A, B)	A + B + Control_input = A + B + 0
1	Subtraction (A, B)	A − B = A + B + Control_Input = A + B + 1

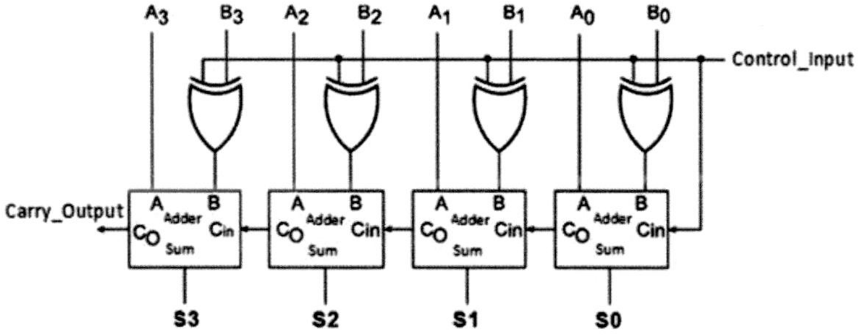

Fig. 3.2 Logic with resource sharing

Logic with Optimized Area: As the full adder is used to perform the addition and subtraction, it eliminates the need of multiplexers and subtractor. The Figure shown is the optimized logic for the 4-bit addition and subtraction. The main advantage of the logic is better data path optimization, and it performs the desired operation depending on the status of the Control_Input (Fig. 3.2).

3.4 Code Converter

Most of the time during the design we may need to use the code converters to convert the data in the suitable format. If we consider the data transfer or control signal transfer in the multiple clock domain designs, then we need to have the gray codes. The section discusses about the binary to gray code converter and gray to binary code converters.

In the two successive binary numbers, one or more than one bit changes, but in the two successive gray codes, only one bit changes. Gray codes are popular and used as pointers in the FIFO design or to pass the control signals in the multiple clock domain designs.

3.4.1 Binary to Gray Code Converter

The 4-bit binary to gray code converter inputs and outputs are listed in Table 3.4.

If we use the input as binary code, then to get the gray code, we need to have the logic as

$$G3 = B3$$
$$G2 = B3 \wedge B2$$
$$G1 = B2 \wedge B1$$

Table 3.4 Binary to gray code

Binary code (B3 B2 B1 B0)	Gray code (G3 G2 G1 G0)
0000	0000
0001	0001
0010	0011
0011	0010
0100	0110
0101	0111
0110	0101
0111	0100
1000	1100
1001	1101
1010	1111
1011	1110
1100	1010
1101	1011
1110	1001
1111	1000

$$G0 = B1 \wedge B0$$

where the character ∧ indicates the XOR operator in the Verilog. The logic diagram is shown below, and the RTL design using Verilog code is described in the example.

Example 1 RTL description of Binary to Gray code converter

```
/////////////////////////////////////////////////////////////////////////////////
module code_converter

            ( input B3, B2, B1, B0,

                  output G3, G2, G1,G0

               );

assign G3=B3;
assign G2 = B3 ^ B2;
assign G1 = B2 ^ B1;
assign G0 = B1 ^ B0;

endmodule
/////////////////////////////////////////////////////////////////////////////////
```

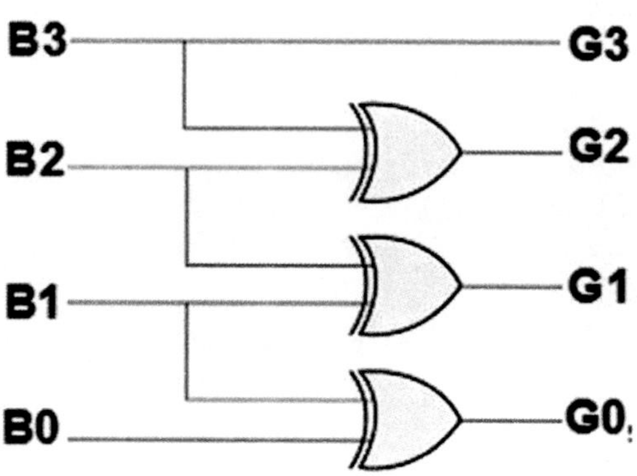

3.4.2 Gray to Binary Code Converter

The 4-bit gray to binary code converter inputs and outputs are listed in Table 3.5.

If we use the input as gray code, then to get the binary code, we need to have the logic as

B3 = G3
B2 = G3 ∧ G2

Table 3.5 Gray to binary code

Gray code (G3 G2 G1 G0)	Binary code (B3 B2 B1 B0)
0000	0000
0001	0001
0011	0010
0010	0011
0110	0100
0111	0101
0101	0110
0100	0111
1100	1000
1101	1001
1111	1010
1110	1011
1010	1100
1011	1101
1001	1110
1000	1111

$$B1 = (G3 \wedge G2 \wedge G1) = B2 \wedge G1$$
$$B0 = (G3 \wedge G2 \wedge G1 \wedge G0) = B1 \wedge G0$$

where the character $\wedge$ indicates the XOR operator in the Verilog. The logic diagram is shown below, and the Verilog RTL is described in the Example.

Example 2 RTL description of Gray to Binary code converter

```
///////////////////////////////////////////////////////////////////////////
module code_converter

                    ( output B3, B2, B1, B0,

                     input G3, G2, G1,G0

                    );

assign B3=G3;
assign B2 = G3 ^ G2;
assign B1 = G3 ^ G2 ^  G1;
assign B0 = G3 ^ G2 ^  G1 ^ G0;

endmodule

///////////////////////////////////////////////////////////////////////////
```

For the FPGA based designs the technology schematic uses the LUTs and shown in Figure (Figs. 3.3 and 3.4).

3.5 Multiplexers

The multiplexers are popular in the ASIC and FPGA design as data selectors, and the multiplexers are treated as universal logic. Using the multiplexers, the desired combinational function of\r logic is implemented. The effective way to infer the multiplexer is using the continuous assign construct with the use of the conditional operator. Consider 2:1 multiplexer (Table 3.6).

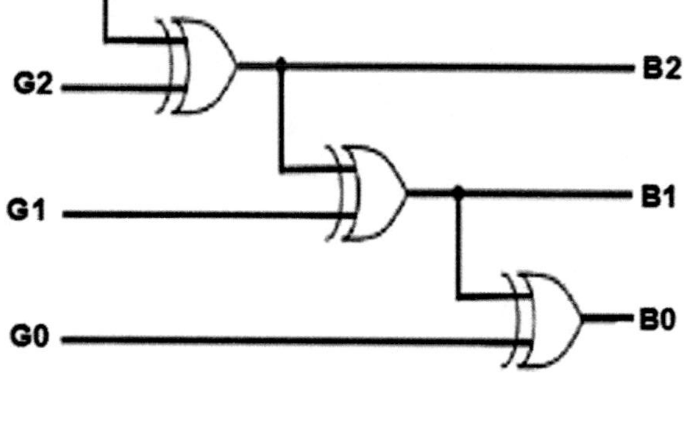

Fig. 3.3 4-bit gray to binary code converter

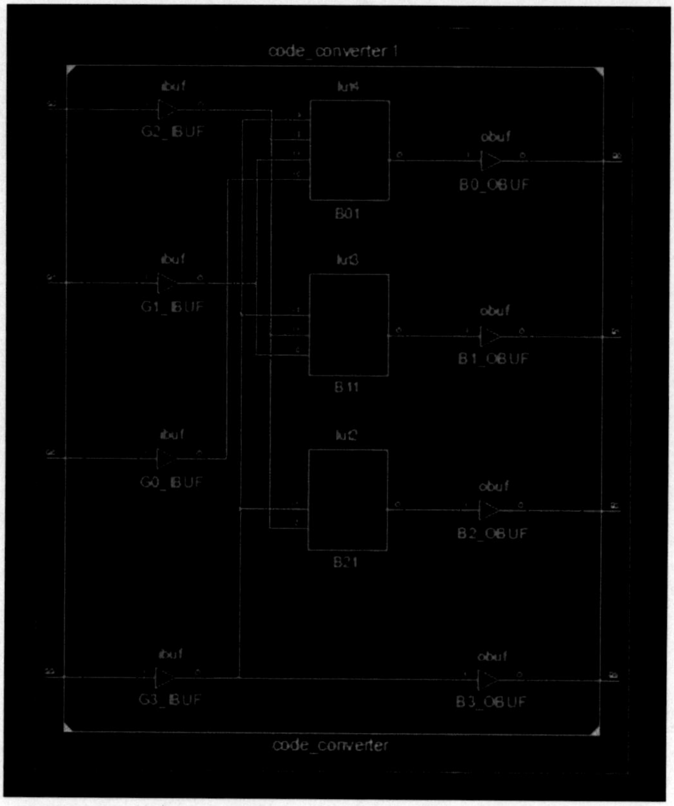

Fig. 3.4 Technology schematic for 4-bit gray to binary code converter

Table 3.6 The 2:1 multiplexer

Control_Input (Sel)	Output (Y)	Description
0	I0	For Sel = 0 an output Y = I0
1	I1	For Sel = 1 an output Y = I1

```
/////////////////////////////////////////////////////////////////////////////////
module mux_2to1( input I1, I0, Sel, output Y );

assign Y = (Sel) ? I1 : I0; // conditional operator is used to infer 2:1 MUX


endmodule

/////////////////////////////////////////////////////////////////////////////////
```

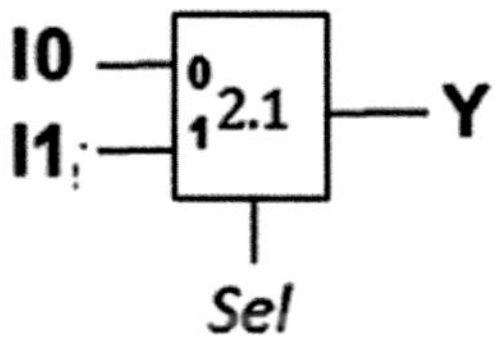

As shown for Sel = 1 Y = I1 and for Sel = 0 Y = I0.

Using the multiple ***assign*** constructs and by using the conditional operator, the 4:1 multiplexer is implemented.

```
///////////////////////////////////////////////////////////////////////////////////
module mux_4to1( input I3, I2, I1, I0,  S1, S0, output Y );

wire Y1, Y0;

assign Y0 = (S0) ? I1 : I0;
assign Y1 = (S0) ? I3 : I2;
assign  Y = (S1) ? Y1 : Y0;


endmodule
///////////////////////////////////////////////////////////////////////////////////
```

The logic inferred is shown in the Figure, and it consists of the three 2:1 multiplexer. The issue with the design is the cascaded stages that is if each multiplexer has 1 ns propagation delay, the logic inferred will have the 2 ns propagation delay.

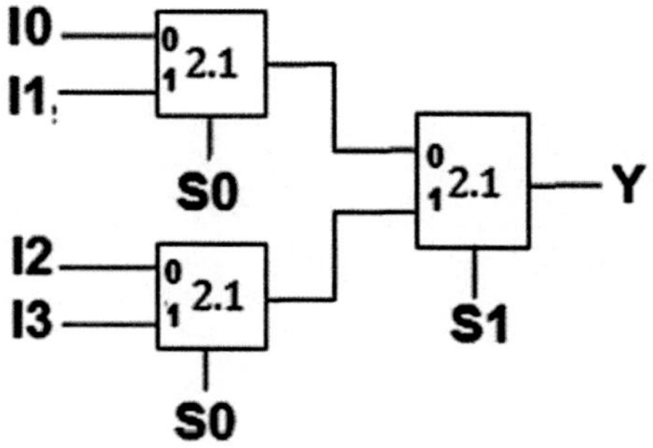

3.6 Cascading Stages of MUX Using Instantiation

```
//////////////////////////////////////////////////////////////////////////////////
module Mux_16to1( input[15:0] I , input [3:0] S, output Y );

wire Y3, Y2, Y1, Y0;

mux_4to1  U0  ( .I3(I[3]),
                .I2(I[2]),
                .I1(I[1]),
                .I0(I[0]),
                .S1(S[1]),
                .S0(S[0]),
                .Y(Y0)
                );

mux_4to1  U1  ( .I3(I[7]),
                .I2(I[6]),
                .I1(I[5]),
                .I0(I[4]),
                .S1(S[1]),
                .S0(S[0]),
                .Y(Y1)

                );
mux_4to1  U2  ( .I3(I[11]),
                .I2(I[10]),
                .I1(I[9]),
                .I0(I[8]),
                .S1(S[1]),
                .S0(S[0]),
                .Y(Y2)
                );
```

```
mux_4to1  U3  ( .I3(I[15]),
                .I2(I[14]),
                .I1(I[13]),
                .I0(I[12]),
                .S1(S[1]),
                .S0(S[0]),
                .Y(Y3)
                );

mux_4to1 U4 ( .I3(Y3),
              .I2(Y2),
              .I1(Y1),
              .I0(Y0),
              .S1(S[3]),
              .S0(S[2]),
              .Y(Y)
              );

endmodule
//////////////////////////////////////////////////////////////////////////////////
```

The logic inferred consists of five four to one multiplexers and is inferred due to the component instantiation.

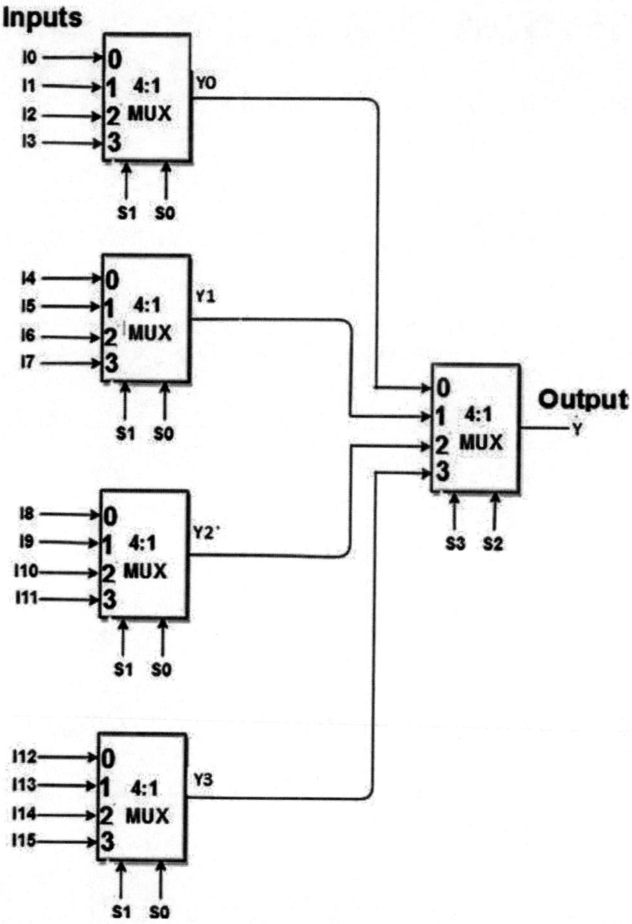

If we consider the ASIC-based designs, then the multiplexers are useful in following tasks.

1. The multiplexers are used to select from many clock frequencies and used as clock selector.
2. Used for the pin multiplexing to minimize the pin count of the design
3. Used as combinational logic to select from one of the data inputs.

3.7 Decoders

The decoders are extensively used in the ASIC- and FPGA-based designs to select one of the devices to establish communication. In the decoders, one of the outputs is active at a time depending on the status of the select inputs.

Consider the 2:4 decoder which has two select inputs and four outputs. The decoder is enabled when Enable = 1. Table 3.7 describes the relationship between the select lines and output lines.

Strategy During RTL Design: To infer the decoders which has 16 or 32 inputs, we can use the logic duplication concept where the 'Enable' input of the decoder can be controlled by the input side decoder. The designer can use the module instantiation or the RTL design using the case construct to infer the required logic (Table 3.8).

Table 3.7 Truth table 2:4 decoder

Enable	Select inputs (I1 I0)	Outputs (Y3 Y2 Y1 Y0)
1	00	0001
1	01	0010
1	10	0100
1	11	1000
0	xx	0000

Table 3.8 Truth table 4:16 decoder

Enable	Select inputs (I3 I2 I1 I0)	Output in binary
1	0000	0000_0000_0000_0001
1	0001	0000_0000_0000_0010
1	0010	0000_0000_0000_0100
1	0011	0000_0000_0000_1000
1	0100	0000_0000_0001_0000
1	0101	0000_0000_0010_0000
1	0110	0000_0000_0100_0000
1	0111	0000_0000_1000_0000
1	1000	0000_0001_0000_0000
1	1001	0000_0010_0000_0000
1	1010	0000_0100_0000_0000
1	1011	0000_1000_0000_0000
1	1100	0001_0000_0000_0000
1	1101	0010_0000_0000_0000
1	1110	0100_0000_0000_0000
1	1111	1000_0000_0000_0000
0	xxxx	0000_0000_0000_0000

Table 3.9 The 2:4 decoder as selector

Enable	Select inputs (I3 I2)	Outputs (Y3 Y2 Y1 Y0)
1	00	Enable Decoder 1
1	01	Enable Decoder 2
1	10	Enable Decoder 3
1	11	Enable Decoder 4
0	xx	Disable all output decoders

If we observe the I3, I2 inputs, then the binary value remains same for the group of consecutive four entries starting from 0 and can be used to design the inputs decoder. The input decoder is used to select one of the output decoders at a time (Table 3.9).

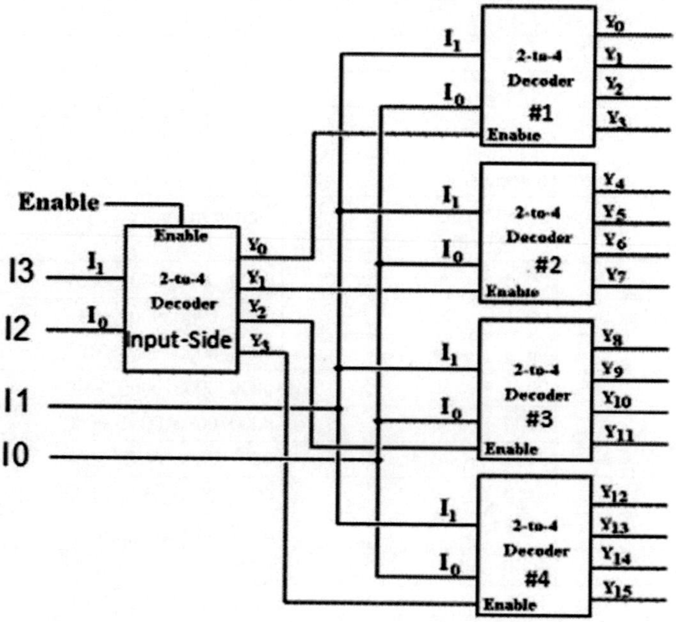

From the figure, it is clear that the input-side decoder is used to enable one of the output decoders at a time. The logic duplication is popular technique and useful in the ASIC and FPGA designs to replicate the same logic. This technique is especially useful in the FPGA-based designs to reduce the number of LUTs. For example, if we need to infer the 8:256 decoder, then using the case, construct the logic inferred in having a greater number of LUTs. So, larger case constructs can be splitted into multiple using the logic duplication techniques. For ASIC, we can think of using the logic duplication depending on the specific scenarios as it impacts on the area of functional block.

Table 3.10 The 4:2 encoder truth table

Select inputs (I3 I2 I1 I0)	Outputs (Y1 Y0)	Status
1000	11	0
0100	10	0
0010	01	0
0001	00	0
0000	00	1

3.8 Encoders

The encoder performs the reverse operation as compared to decoder. Consider the four level-sensitive inputs and the design need to generate an output depending on the logic high value at one of the inputs.

If I3 = 1 and I2, I1 and I0 are logic 0, then the encoder generates valid output as Y1 = 1 and Y0 = 1. As outputs are valid, the status output is 0.

If all inputs are logic zero, then Status = 1 and indicates the invalid output and should be ignored by other design which uses encoder outputs as inputs.

As shown in Table 3.10, it is assumed that one of the inputs is logic '1' at a time, but in practical systems, the assumption does not hold good as many inputs can be active high at a time. So, to cater the need of the priority scheduling, it is essential to design the priority encoders.

3.9 Priority Encoders

The priority encoders are used in the design to sample the high priority input and to generate an valid output.

Consider the design having four inputs I3, I2, I1, I0, and I3 has highest priority as compared to other inputs. Table 3.11 describes the functionality where I3 has highest priority and I0 has lowest priority.

Table 3.11 The 4:2 priority encoder truth table

Select inputs (I3 I2 I1 I0)	Outputs (Y1 Y0)	Status
1xxx	11	0
01xx	10	0
001x	01	0
0001	00	0
0000	00	1

The RTL should be coded using the nested if-else construct and should infer the priority logic.

```
/////////////////////////////////////////////////////////////////////////////////////


module  priority_encoder (
                    input    I3, I2, I1, I0,

                   output  reg  Y1, Y0, flag
                         );

always @ *
begin

 if ( I3)
      begin
       Y1 = 1;
      Y0=1;
      flag = 0;
      end
      else if ( I2)
      begin
       Y1 = 1;
      Y0=0;
      flag=0;
      end
       else if ( I1)
      begin
      Y1 = 0;
```

```
        Y0=1;
        flag=0;
        end
         else if (I0)
       begin
        Y1 = 0;
        Y0=0;
        flag=0;
        end

        else
         begin
        Y1 = 0;
        Y0=0;
        flag=1;
        end

end

endmodule

/////////////////////////////////////////////////////////////////////////////////
```

The FPGA synthesis is shown in the figure, and the technology schematic consists of three lookup tables. FPGA important resource is CLB, and it consists of the LUTs and slice registers.

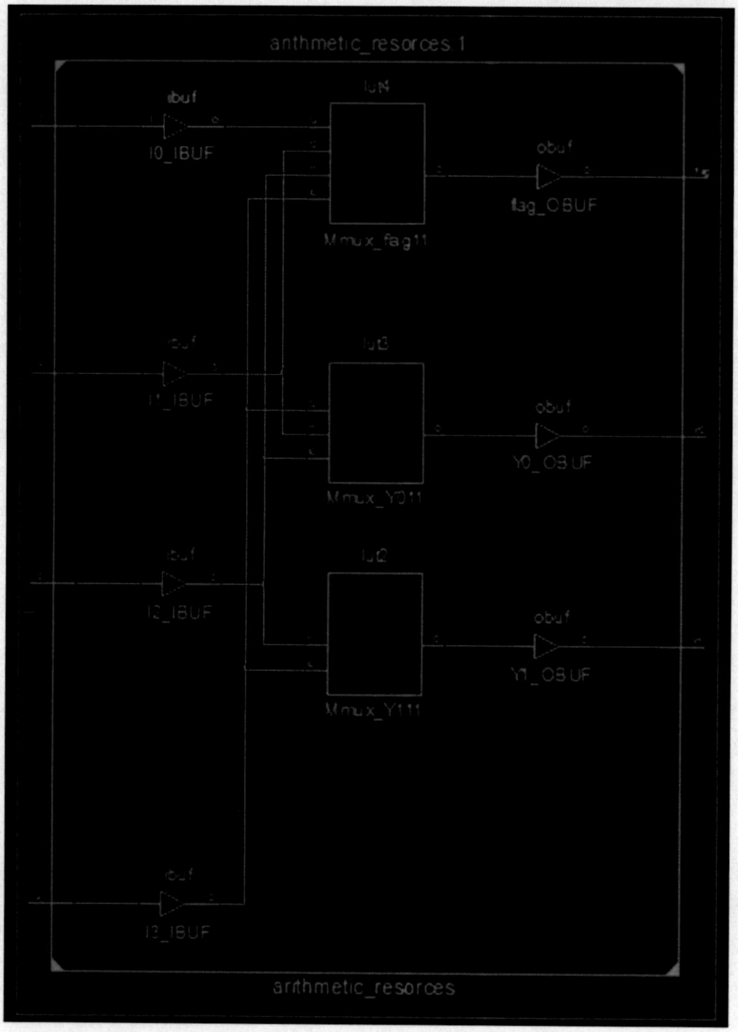

3.10 Strategies During ASIC Design

During the ASIC design flow at the RTL design phase, try to use the following strategies.

1. Use the synthesizable constructs and do not use the #delays in the RTL design.
2. Try to avoid the use of multiple assign constructs as due to parallel execution the area may be higher.

3. Try to use ***always*** @ * to model the combinational logic. It includes the required inputs in the sensitivity list.
4. Use the blocking assignments to code the RTL using the always procedural block.
5. Try to use the naming conventions such as net_name_in, net_name_out for better readability.
6. Try to avoid the combinational looping as they have oscillatory behavior.
7. Try to use the parameter to have parameterized design.
8. Use the ***case*** construct to infer the parallel logic and nested ***if-else*** to infer the priority logic.
9. While using ***casez*** and ***casex*** take care of the dangling inputs and synthesis and simulation mismatches.
10. Try to divide the large number of assignments using multiple ***case*** constructs.

3.11 Exercises

1. Implement the digital logic to detect the even parity from the string of 8-bit binary data? What should be the RTL design strategy for ASIC design?
2. Work on the logic development of 4-bit multiplier. Find out the resource requirement for the 4-bit multiplier for the shift and add method.
3. If propagation delay of the full adder is 1 ns to generate the sum and 2 ns to generate the carry output, and if the 16-bit ripple carry adder is to be designed, then find the overall propagation delay to get the add (A, B)
4. Can you think of the high-level design strategy to pass data between the pipelined processor and the memory and IOs. If the four memory blocks of 16 Kbyte each and 128 IO devices need to establish communication with the processor, then what should be the decoding logic?
5. Sketch the logic and optimize the area required for following

   (a) A OR B
   (b) A AND B
   (c) A XOR B
   (d) NOT A

Consider A, B are 8-bit binary inputs, and to perform only one operation at a time, the design has 2-bit control inputs (control_operation).

3.12 Chapter Summary

Following are important points to conclude the chapter.

1. Use the resource sharing techniques to optimize the logic for the better data path optimization.

2. The multiplexer is universal logic, and the chain of multiplexers can be used to select from one of the inputs.
3. Multiplexers are used to have pin multiplexing.
4. Decoders are used as selection logic in the system design.
5. Priority encoders are used to generate the output by analysing the priority of inputs.

Chapter 4
Sequential Design Concepts

If we consider the ASIC design cycle then at various stages, we need to play around with the sequential circuits may be using the counters, shift registers, memories and other clock-based circuits and clock dividers. In such scenario, we may encounter few issues where we need to tweak the RTL or architecture to improve the performance of ASIC. By considering all the above points, the chapter discusses about important sequential elements with their use and the design of synchronous and asynchronous sequential circuits.

4.1 Sequential Design Elements

In the sequential circuits, an output is function of the present input and past output. Table 4.1 discuss about the important sequential elements and their use during the design phase. For the synchronous or asynchronous designs, we will try to use the edge-sensitive elements that are D flip-flops and our objective is to design the sequential circuit with speed as the timing goal.

Understanding of the synchronous and asynchronous design techniques plays important role during the design cycle.

V. Taraate, *ASIC Design and Synthesis*,
https://doi.org/10.1007/978-981-33-4642-0_4

Table 4.1 Important sequential elements

Sequential elements	Description
Latch	Level sensitive and used in the latch-based designs. Latch-based clock gating and few scenarios in DFT for time borrowing
Flip-flop	Edge sensitive and mainly used in the design of asynchronous and synchronous sequential circuits. Application can be counters, shifters, design with the registered inputs and registered outputs
Counters	The synchronous or asynchronous counters where counting happens on the active edge of the clock. Used to count the predefined sequence and also used as clock dividers
Shift-registers	Sensitive on the active edge of clocks and designed using D flip-flops and used to perform the right, left shift and rotate operations
Asynchronous circuits	Can be used to generate clocks but incurs more delay so avoid the use of the asynchronous circuits
Clock dividers	The PLL clock can be divided to get the internally generated clock using the D flip-flops (Toggle flip-flop designed using D flip-flop)
Finite state machine (FSM) controller	For the larger ASIC designs different FSM controllers can be used to detect the sequence with the glitch free and clean output
Random number generators	Used to generate the random numbers for specific requirements using edge-sensitive elements and suitable combinational elements

4.2 Let Us Understand Blocking Versus Non-blocking Assignments

Important Verilog assignments used within the procedural block 'always' and 'initial' are

- Blocking Assignments (BA)
- Non_Blocking Assignments (NBA).

The BA assignments are evaluated and updated in the active event queue, and the right-hand side of the NBA is evaluated in the active event queue and updated in the NBA queue.

The name blocking indicates that it blocks all the future assignments unless and until present assignment is evaluated and updated, and hence, BA are not recommended to use in the sequential design. It is recommended to use the BA for the combinational design.

4.2.1 Blocking Assignments

The RTL description using the blocking assignments is shown in Example 1 and the intent is to infer the 3-bit shift register. But it infers the single flip-flop as the future assignment evaluation is blocked during the current assignment execution.

Example 1 RTL description using the blocking assignments

```
////////////////////////////////////////////////////////////////////////////////

module blocking_assignments( input data_in, clk, reset_n, output reg
data_out);

reg [1:0] tmp;

always @ ( posedge clk or negedge reset_n)
begin

        if ( ~ reset_n)
        begin
                { data_out, tmp } = 3'b000;

        end

        else
  begin
                tmp[0] = data_in;
                tmp[1] = tmp[0];
                data_out = tmp[1];

         end
end
endmodule
////////////////////////////////////////////////////////////////////////////////
```

The RTL schematic is shown in Fig. 4.1 and the inferred logic has single flip-flop with asynchronous reset.

4.2.2 Reordering of the Blocking Assignments

If we reorder the blocking assignments as shown in Example 2, the RTL infers the 3-bit shift register, and it indicates that the order of the blocking assignments will play important role during the RTL description.

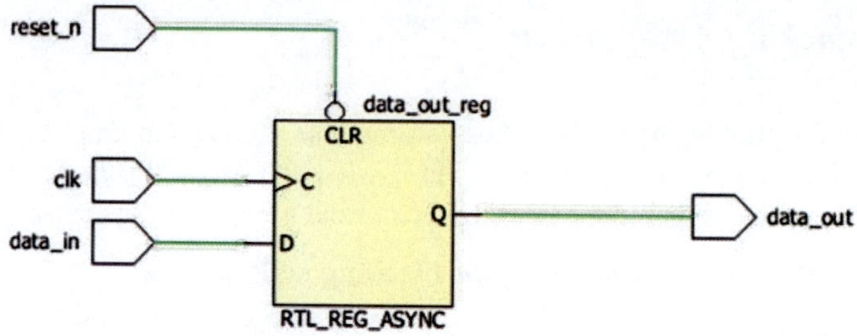

Fig. 4.1 RTL schematic for Example 1

Example 2 RTL description using the blocking assignments reorders

```
//////////////////////////////////////////////////////////////////////

module blocking_assignments( input data_in, clk, reset_n, output reg
data_out);

reg [1:0] tmp;

always @ ( posedge clk or negedge reset_n)
begin

        if ( ~ reset_n)
        begin
                { data_out, tmp } = 3'b000;

        end

        else
   begin
                data_out = tmp[1];
                tmp[1] = tmp[0];
                tmp[0] = data_in;

                end
end
endmodule

//////////////////////////////////////////////////////////////////////
```

The RTL schematic is shown in Fig. 4.2, and the RTL design description (Example) infers the 3-bit shift register using the asynchronous active low reset.

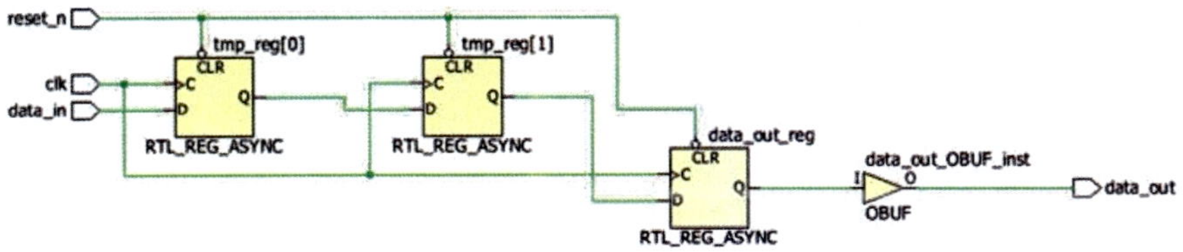

Fig. 4.2 RTL schematic for Example 2

4.2.3 Non-blocking Assignments

The RTL description using the non-blocking assignments is shown in Example 3 and the intent is to infer the 2-bit shift register, and it infers the 2-bit shifter using D flip-flops as the assignments executes concurrently within the ***begin-end***.

Example 3 RTL description using the non-blocking assignments

```
/////////////////////////////////////////////////////////////////////////
module non_blocking_assignments( input data_in, clk, reset_n, output
reg data_out);

reg tmp;

always @ ( posedge clk or negedge reset_n)
begin

        if ( ~ reset_n)
        begin
                { data_out, tmp } <= 2'b00;

        end

        else
                begin

                        data_out <= tmp;

                        tmp <= data_in;
                end
end
endmodule

/////////////////////////////////////////////////////////////////////////
```

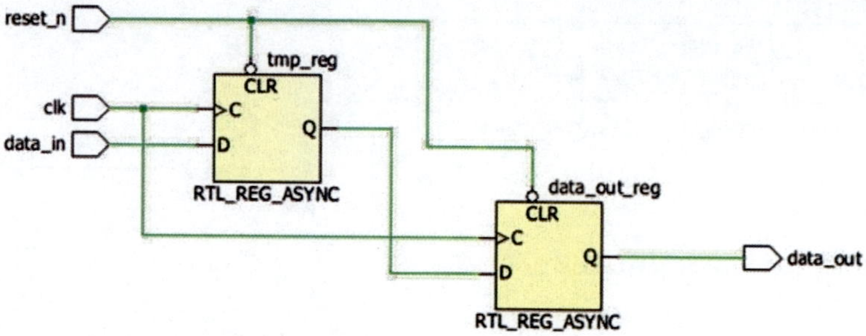

Fig. 4.3 RTL schematic for Example 3

The RTL schematic is shown in Fig. 4.3, and the RTL design description (Example 3) infers the 2-bit shift register using the asynchronous active low reset.

4.2.4 Reordering of the Non-blocking Assignments

Reordering of the non-blocking assignments as shown in Example 4, the RTL synthesis infers the 2-bit shift register, and it indicates that the order of the non-blocking assignments will not effect on the synthesis result of the design. The RTL schematic is shown in Fig. 4.3.

Example 4 RTL description using the non-blocking assignments reorders

```
/////////////////////////////////////////////////////////////////////////////
module non_blocking_assignments( input data_in, clk, reset_n, output
reg data_out);

reg tmp;

always @ ( posedge clk or negedge reset_n)
begin

        if ( ~ reset_n)
        begin
                { data_out, tmp } <= 2'b00;

        end

        else
                begin

                        data_out <= tmp;

                        tmp <= data_in;
                end
end
endmodule
/////////////////////////////////////////////////////////////////////////////
```

Use the following recommendations during the ASIC RTL designs

1. Use the non-blocking assignments to model the sequential logic
2. Use the blocking assignments to model the combinational logic
3. Don't mix blocking and non-blocking assignments.

4.3 Latch-Based Designs

Latches are level sensitive and used in few applications such as latch-based clock gating in low-power ASIC designs and few other applications such as DFT scan chains with the latch borrowing.

Consider the functionality of the positive-level sensitive latch where the clk acts as latch enable input. When the latch enable is active high, the latch acts as transparent and Q = D where Q is an output of latch and D is data input. When latch is not enabled that is clk = 0, the latch holds previous value as previous output Q is circulated through the even number of NOT gates.

Consider the CMOS implementation of positive-level sensitive latch where the CMOS switch-1 passes data for CLK = 1 to generate valid output Q = D. Latch is transparent during the positive level of the CLK. The even number of NOT gates is used in the forward path and during the negative level of CLK that is CLK = 0 the CMOS switch-2 is ON which holds the previous output Q.

So for the better understanding, the functionality is described in Table 4.2 (Fig. 4.4).

The timing sequence for the positive-level sensitive latch is shown in Fig. 4.5 and the latch output Q = D during the positive level of the CLK. During negative level of clk, it holds the previous output value.

RTL Design Strategy

The Intentional latches can be inferred if during the RTL design else clause is eliminated. Consider the 4-bit latch coded using the Verilog using the if-else construct within the always @ * procedural block. As discussed in Chap. 3 'if-else' generates the 2:1 multiplexer. Now to infer the intentional latches eliminate the 'else' condition.

Table 4.2 The D latch truth table

CLK = Enable	D	Q	CMOS switch status
1	0	0	CMOS Switch-1 is ON and CMOS Switch-2 is OFF. Q = D
1	1	1	CMOS Switch-1 is ON and CMOS Switch-2 is OFF. Q = D
0	X	Hold past output	CMOS Switch-1 is OFF and CMOS Switch-2 is ON. Q = Previous Output

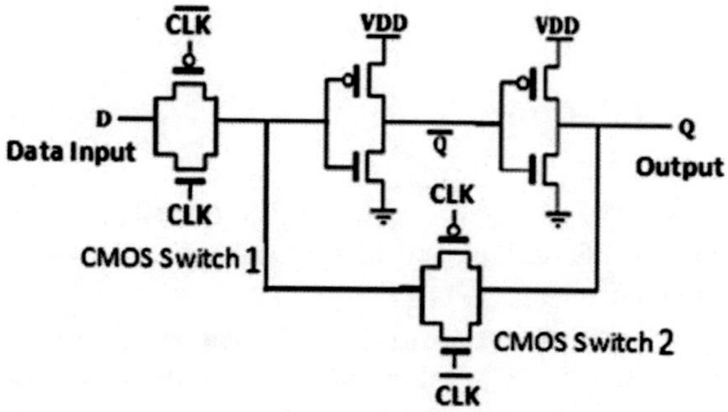

Fig. 4.4 D latch

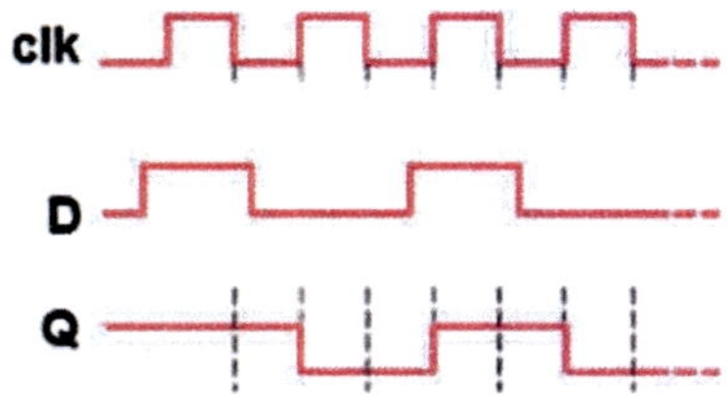

Fig. 4.5 Timing diagram of the D latch

During the ASIC or FPGA synthesis you will get the warning that: Latches may be generated from incomplete case or if statements. We do not recommend the use of latches in designs, as they may lead to timing problems.

Example 5 is RTL description of 4-bit latch and coded using the Verilog synthesizable constructs.

Example 5 RTL description of 4-bit latch

```
/////////////////////////////////////////////////////////////////////////////
module latch_4bit (
                    input [3:0] d_in,
                    input clk,
                   output reg [3:0] q_out
                    );

always @ *

begin

  if (clk)

          q_out <= d_in;

end

endmodule

/////////////////////////////////////////////////////////////////////////////
```

The four-bit latch schematic is shown in Fig. 4.6, and latch is level sensitive as D is sampled during the active high level of CLK.

Fig. 4.6 RTL schematic for Example 5

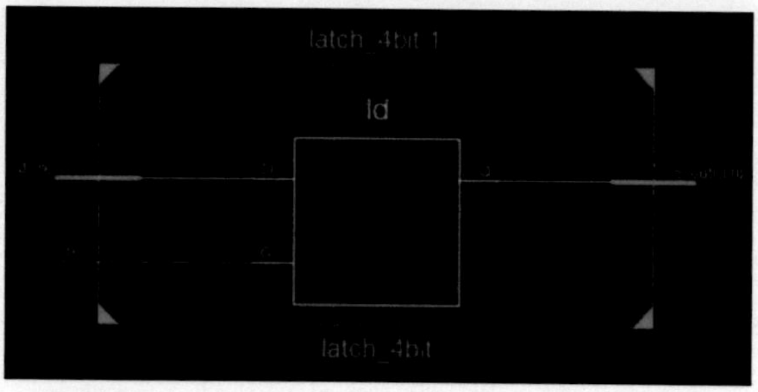

4.4 Flip-Flop-Based Designs

The flip-flops are edge sensitive; that is, they either operate on the rising edge (low to high transition) or falling edge (high to low transition) and used as sequential design element in the synchronous or asynchronous designs.

The falling edge D flip-flop is shown in Fig. 4.7, and on the high to low transition of the clk, the flip-flop samples the data input D to generate valid output.

The functional Table 4.3 describes the relationship between input and output on the rising edge of clk.

The latch based implementation of the negative edge-sensitive flip-flop is shown in Fig. 4.7. As shown, the flip-flop has two latches connected in cascade. The master latch is positive-level sensitive, and the slave latch is negative-level sensitive.

The timing sequence for the negative edge-sensitive flip-flop is shown in the Fig. 4.8. As shown, the master latch output Q1 changes on the positive level of clk and slave latch output Q changes on the negative level of clock. The output from flip-flop Q is sensitive to the falling edge of clk.

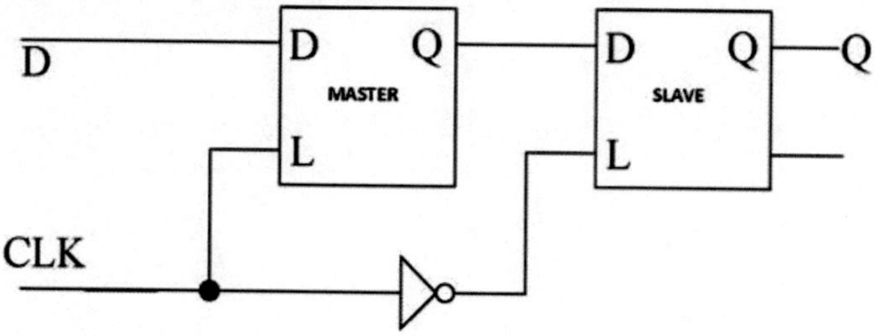

Fig. 4.7 D flip-flop

Table 4.3 Truth-table of negative edge-triggered flip-flop

CLK	D	Q
Falling edge	0	0
Falling edge	1	1
Rising edge or level	X	Hold past output

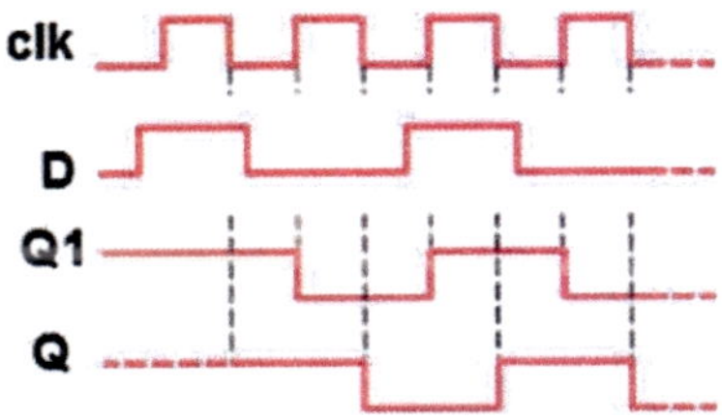

Fig. 4.8 Timing diagram of D flip-flop

> The flip-flop or parallel-in-parallel-out registers are used in the ASIC design to hold the data or to change the data on the active edge of the clock.

> The edge sensitive elements are inferred if we use the procedural block ***always*** with event control as ***posedge*** or ***negedge***.

> To infer the edge sensitive element use either one from following
> ***always @ (posedge clk),***
> ***always @ (negedge clk)***

Example 6 is the RTL description of the 4-bit parallel-in-parallel-out register. The schematic is the representation of the 4-bit register which is rising edge-triggered and uses four flip-flops (Fig. 4.9).

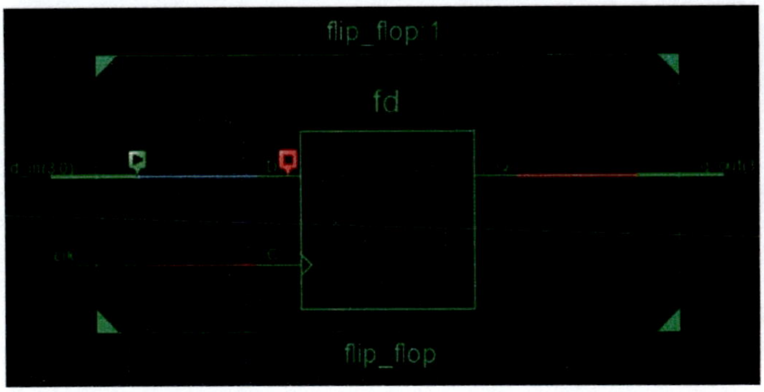

Fig. 4.9 RTL schematic for Example 6

Example 6 RTL description of 4-bit parallel-input-parallel-output register

```
////////////////////////////////////////////////////////////////////////////
module  flip_flop  (

                            input [3:0] d_in,
                            input clk,
                            output reg [3:0] q_out
                       );

always @ (posedge clk)

begin

   q_out <= d_in;


end

endmodule

////////////////////////////////////////////////////////////////////////////
```

4.5 Reset Strategies

For the FPGA and ASIC designs, the role of reset to initialize the design plays an important rule. During the ASIC implementation to synchronize the internal resets with the master reset input at various stages, we need to deploy the reset synchronizers.

The dedicated reset tree and reset network can be used to generate the reset signal required for the different functional blocks.

For the reset, we need to consider the

- Reset Recovery Time
- Reset Removal Time.

The resets are of two types.

1. **Asynchronous Reset**
2. **Synchronous Reset**.

4.5.1 Asynchronous Reset

If the reset need to be used to initialize the design irrespective of the active edge of the clock, then the reset is treated as asynchronous reset. The ASIC RTL design which uses the D flip-flop is shown in Example 7.

Example 7 RTL description of the asynchronous-reset D flip-flop

```
///////////////////////////////////////////////////////////////////////////
module flip_flop(

                input  d_in,
                 input clk, reset_n,
                 output reg q_out

                                         );

always @ (posedge clk or negedge reset_n)

begin

        if ( ~ reset_n)
        q_out <= 1'b0;

        else

       q_out <= d_in;


end
endmodule

///////////////////////////////////////////////////////////////////////////
```

The reset is asynchronous input (reset_n) and the RTL schematic is shown in Fig. 4.10.

The testbench for Example 8 is described using the non-synthesizable constructs and is shown in example, and the sanity check result is shown in the waveform (Fig. 4.11).

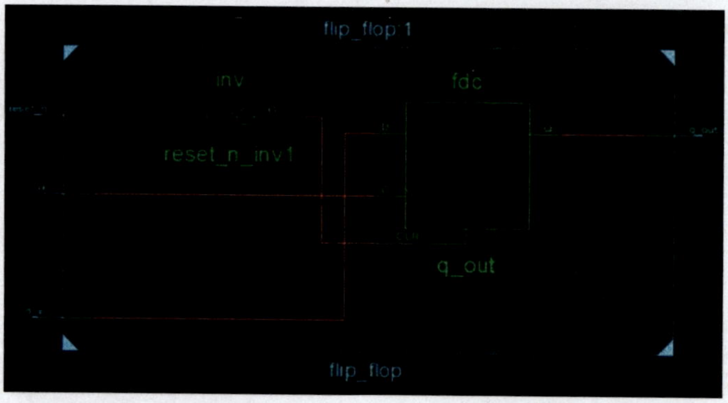

Fig. 4.10 Asynchronous-reset logic

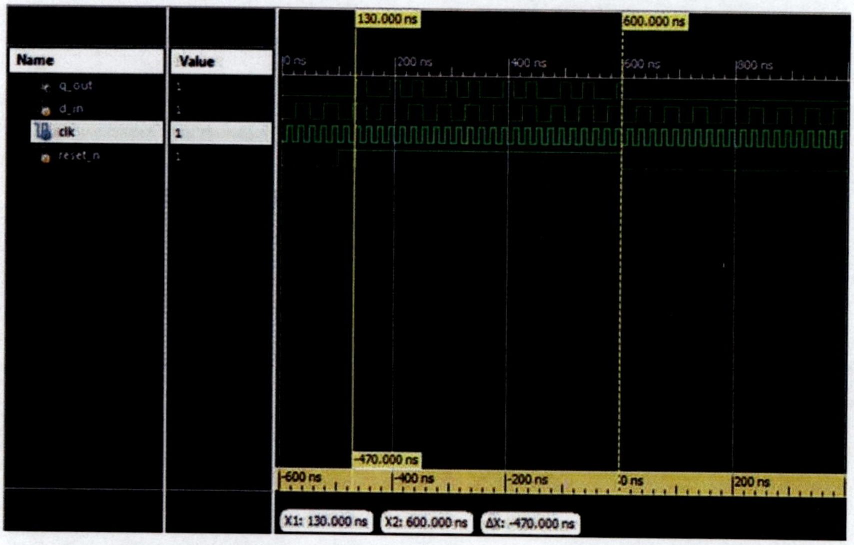

Fig. 4.11 Asynchronous-reset sampling

Example 8 Testbench for the D flip-flop

```
//////////////////////////////////////////////////////////////////////////////

module test_flip_flop;

        // Inputs
        reg d_in;
        reg clk;
        reg reset_n;
        // Outputs
        wire q_out;

        // Instantiate the Unit Under Test (UUT)
                flip_flop  uut  (
                .d_in(d_in),
                .clk(clk),
                .reset_n(reset_n),
                .q_out(q_out)
        );

always #10 clk=~clk;

// Add stimulus here

always # 25 d_in = ~d_in;

        initial begin
                // Initialize Inputs
                d_in = 0;
                clk = 0;
                reset_n = 0;

                // Wait 100 ns for global reset to finish
                #100;

                reset_n =1'b1;


                #500 reset_n = 1'b0;

                #400 $finish;


        end
endmodule

//////////////////////////////////////////////////////////////////////////////
```

As shown in the simulation waveform the reset is sampled irrespective of the active edge of clock and used to initialize the state of flip-flop to ‘0’.

4.5.2 Synchronous Reset

If the reset need to be used to initialize the design on, the active edge of the clock then the reset is treated as synchronous reset. The ASIC RTL design which uses the D flip-flop is shown in Example 9.

Example 9 RTL description of the synchronous-reset D flip-flop

```
///////////////////////////////////////////////////////////////////////////
module flip_flop (

                input d_in,
                input clk, reset_n,
                output reg q_out

               );

always @ ( posedge clk )

begin

        if ( ~ reset_n)

        q_out <= 1'b0;

        else

        q_out <= d_in;

end

endmodule
///////////////////////////////////////////////////////////////////////////
```

As the always procedural block is sensitive to positive edge of the clock, the reset is checked within the always procedural block and it reset_n = 0; the output of the flip-flop is initialized to logic '0'.

The testbench for Example 9 is described using the non-synthesizable constructs and is shown in Example 8, and the sanity check result is shown in Fig. 4.12.

As shown in the simulation waveform, the reset is sampled on the active edge of clock and used to initialize the state of flip-flop to '0'.

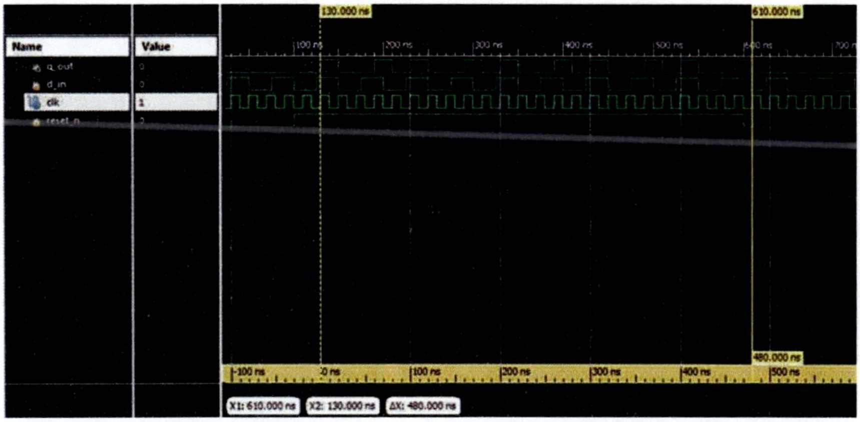

Fig. 4.12 Synchronous-reset sampling

4.6 Frequency Divider

Let us consider the use of the flip-flop in the ASIC or FPGA designs. We can use the D flip-flops with additional combinational logic if required to design the clock divider.

Consider that the PLL generates the 550 MHz square wave and used as a clock. For internal logic, we need to use the 225 MHz clock; then in such scenario, we can design the divide-by-two circuits. The circuit can be designed using the D flip-flop where the output complement of Q is feedback to the input D (Fig. 4.13).

Example 10 RTL description of the clock divider logic

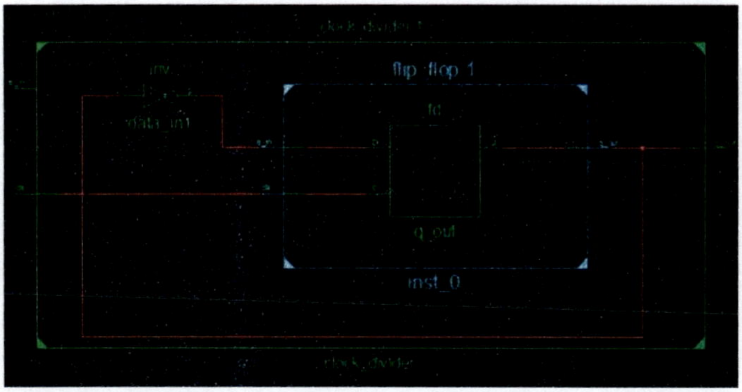

Fig. 4.13 Clock divider logic

```
///////////////////////////////////////////////////////////////////////////////
module clock_divider (
                              input d_in,
                              input clk,reset_n,
                              output q_out

                                                          );


wire data_in;
flip_flop inst_0 (

                    .d_in(data_in),
                    .clk(clk),
                    .reset_n(reset_n),
                    .q_out(q_out)

                    );

assign data_in = ~q_out;
endmodule


///////////////////////////////////////////////////////////////////////////////
```

As shown in Example 11 for the divide-by-two circuits is described using the non-synthesizable constructs. The simulation waveform for the design is shown in the Fig. 4.14. The synthesis outcome is shown in the Fig. 4.15.

Example 11 Testbench to check for the clock division

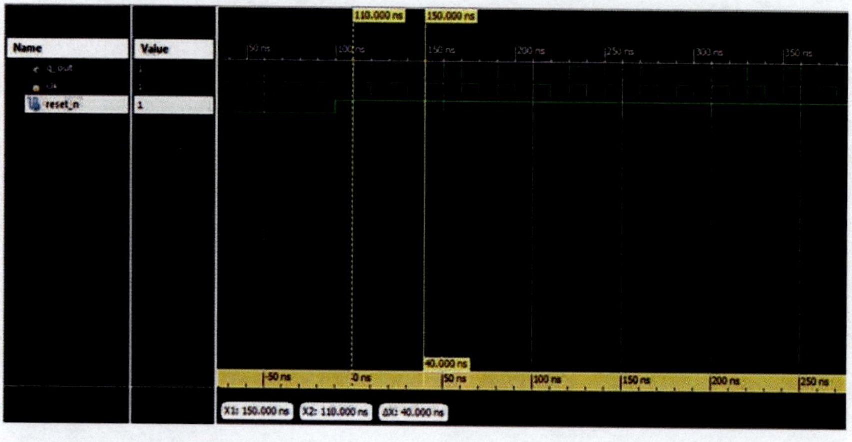

Fig. 4.14 Waveform for Example 11

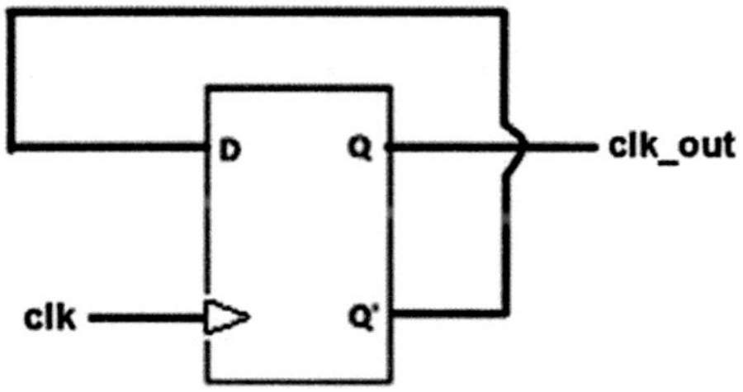

Fig. 4.15 ASIC divide-by-two clock logic

```
/////////////////////////////////////////////////////////////////////////////////

module divide_by_2_test;

        // Inputs
        reg clk;
        reg reset_n;

        // Outputs
        wire q_out ;

        // Instantiate the Unit Under Test (UUT)
        clock_divider   uut (
                                .clk(clk),
                                .reset_n(reset_n),
                                .q_out(q_out)
                        );

        initial begin
                // Initialize Inputs
                clk = 0;

                reset_n =1'b0;

                #100;
```

```
        // Wait 100 ns for global reset to finish

        reset_n = 1'b1;

        #500 reset_n = 1'b0;


        end

    always # 10 clk = ~ clk;

endmodule
```

Issue of using above RTL during ASIC Design: With above Verilog RTL we can get the divide by 2 sequential circuits which are sensitive to rising edge of the clock. But the design has issue due to use of the NOT gate as it increases the data required time. This limits the maximum frequency of the design.

How to fix the Issue: To improve the maximum frequency for the clock divider use fed the complement of Q from the flip-flop directly to data input D. For more details refer Chaps. 5 and 10.

4.7 Synchronous Design

In the synchronous design the all sequential element that is D flip-flops uses the clock generated from the common clock source. For the better performance of the ASIC and to have the clean timing, it is recommended to use the synchronous design.

The 2-bit binary up-counter which is used to count the sequence 00-01-10-11-00…. is described in the RTL (Example 12) using the synthesizable constructs. The RTL design uses the asynchronous active low reset.

Example 12 RTL description of the 2-bit synchronous counter

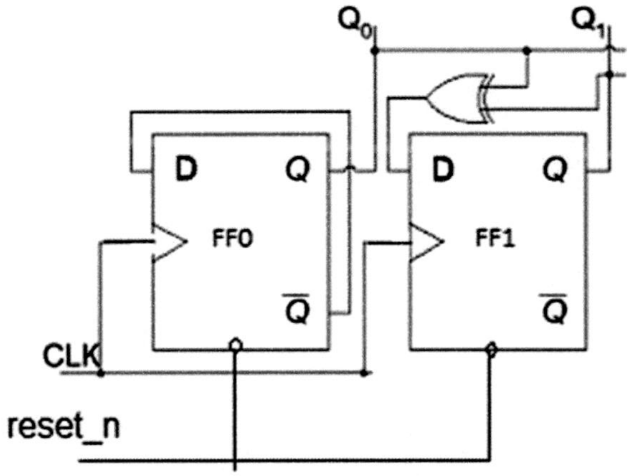

Fig. 4.16 2-bit synchronous binary up-counter schematic

```
////////////////////////////////////////////////////////////////////////////
module Binary_up_counter (
                        input clk,reset_n,
                        output reg [1:0] Q
                        );

always @ (posedge  clk or negedge  reset_n)
begin

 if (~ reset_n)

        Q <= 2'b00;

else

  if ( Q== 2'b11)

          Q <= 2'b00;

  else

          Q <= Q+1;
end
endmodule
////////////////////////////////////////////////////////////////////////////
```

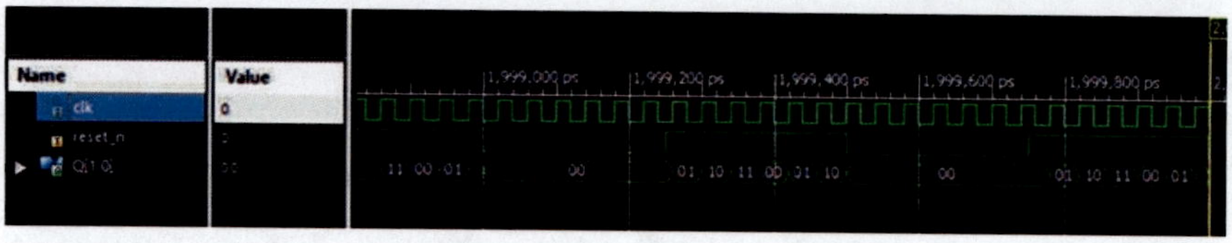

Fig. 4.17 Timing diagram of the 2-bit binary up-counter

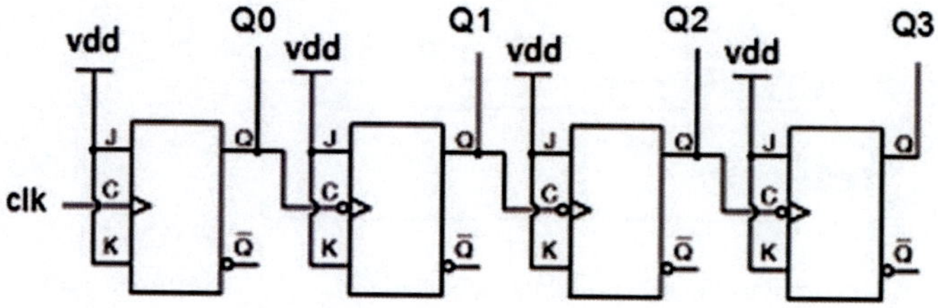

Fig. 4.18 Asynchronous counter as clock divider

The RTL description (Example 12) infers the logic and shown in Fig. 4.16. The simulation waveform for the 2-bit counter is shown in Fig. 4.17.

4.8 Asynchronous Design

In the asynchronous design the clock of all the sequential elements is not from the common clock source. As shown in Fig. 4.18, the design uses the rising edge-sensitive toggle flip-flops, and the design is used to generate the internal clocks.

If the clock frequency at clk input is 500 MHz, then the internally generated clocks are the following

- Clock output Q0 having frequency of the 250 MHz.
- Clock output Q1 having frequency of the 125 MHz.
- Clock output Q2 having frequency of the 62.5 MHz.
- Clock output Q3 having frequency of the 31.25 MHz.

The issue with the design is the asynchronous clocking, and hence, it is very dangerous to use such kind of designs in the ASIC.

4.9 RTL Design and Verification for Complex Designs

For the complex designs during the RTL design and verification phase, use the following strategies.

1. Try to understand the architecture and micro-architecture and partition the logic to have the efficient RTL description using the moderate gate count blocks.
2. Use the bottom-up approach and during the top-level design try to deploy the synchronizers if required.
3. Use the synthesizable constructs during the RTL design and non-synthesizable constructs during the RTL verification.
4. Use the blocking assignment to model the combinational logic (glue logic between the registers) and non-blocking assignments for the sequential designs
5. Do not mix the blocking and non-blocking assignments.
6. Use the optimization constraints at the RTL level to improve the performance. Refer the subsequent chapters to have better understanding of the design and optimization.
7. Have the better verification architecture and the verification planning for the design.
8. Understand the coverage requirements and work on the verification strategies to achieve the specified coverage goals.

The discussion on the RTL verification topic is not an objective of the book. The subsequent chapters of this book are useful to understand the ASIC design terminology and concepts.

4.10 Exercises

1. Implement the RTL using the synthesizable Verilog constructs to infer following logic (Fig. 4.19).

2. Implement the RTL using the synthesizable Verilog constructs to infer following logic. Use asynchronous active low reset and positive edge-sensitive elements (Fig. 4.20).

3. Implement the RTL using the synthesizable Verilog constructs to infer following logic (Fig. 4.21).

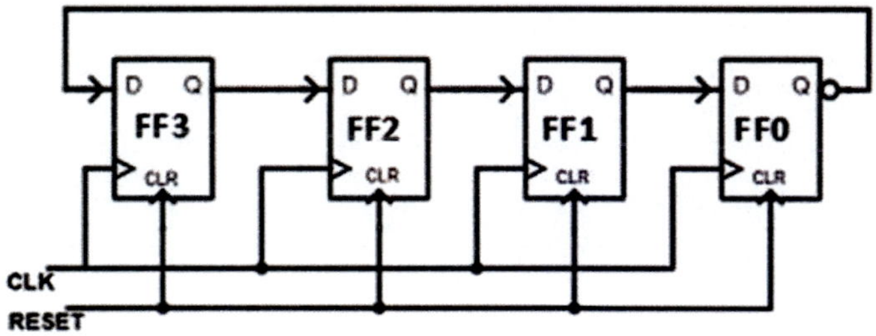

Fig. 4.19 Exercise 1

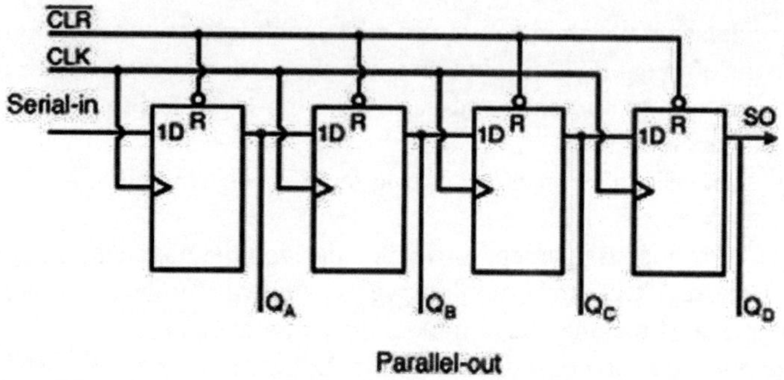

Fig. 4.20 Exercise 2

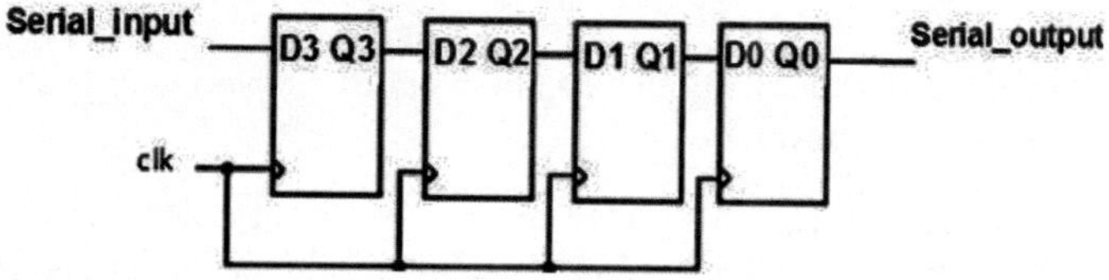

Fig. 4.21 Exercise 3

4.11 Chapter Summary

The following are the important points to conclude the chapter.

1. Use the non-blocking assignments to model the sequential logic.
2. Use the blocking assignments to model the combinational logic.
3. Do not mix blocking and non-blocking assignments.
4. The resets are of the two types asynchronous: reset and synchronous reset.
5. In the synchronous design, the all sequential element that is D flip-flops uses the clock generated from the common clock source.
6. In the asynchronous design, the clock of all the sequential elements is not from the common clock source.
7. Avoid the internally generated clocks or asynchronous clocking in the design.

Chapter 5
Important Design Considerations

If we use the synchronous sequential designs or any IP in the design or to finalize the architecture and micro-architecture, then we need to work out on various strategies. Few of them are listed below:

1. Functionality of the design and compatibility
2. Parallelism, concurrency, and pipelining strategies
3. External IO and high-speed interfaces
4. Area and initial gate count estimation for the design
5. Speed and maximum frequency requirements
6. Power requirements and use of the low-power design cells
7. Clock network and latency
8. Interface and IO delays and modeling strategies
9. Effect of the on-chip variation on the design
10. IPs required and timing requirements
11. Memory requirements and different micros
12. Top and block-level design constraints.

By considering all above, the team of experienced technical members finalizes the architecture and micro-architecture of the ASIC/SOC design.

The chapter discusses few of the design considerations for ease of understanding of the architecture and the case studies.

During the architecture and micro-architecture design, important considerations are speed, power, and area.

V. Taraate, *ASIC Design and Synthesis*,
https://doi.org/10.1007/978-981-33-4642-0_5

5.1 Timing Parameters

Important sequential circuit timing parameters are shown in Fig. 5.1 for rising edge-sensitive flip-flop, and they are:

1. Setup time (t_{su})
2. Hold time (t_h)
3. Propagation delay of flip-flop (t_{pd}).

These timing parameters for the flip-flop are discussed in this section.

- **Setup Time (t_{su})**: The minimum amount of time for which the data input of the flip-flop should maintain the stable value prior to arrival of the active edge of the clock is called as setup time.
  Here, the active edge indicates the low to high transition for rising edge (positive edge)-sensitive flip-flop, and the high to low transition for falling edge (negative edge)-sensitive D flip-flop.
  During the setup time window if the data input changes either from 1 to 0 or vice versa, then the flip-flop output will be metastable which indicates the setup violation. For more details, refer Chaps. 12 and 15.
- **Hold Time (t_h)**: The minimum amount of time for which the data input of the flip-flop should maintain the stable value after the arrival of the active edge of the clock is called as hold time.
  Here, the active edge indicates the low to high transition for rising edge (positive edge)-sensitive flip-flop, and the high to low transition for falling edge (negative edge)-sensitive D flip-flop.
  During the hold time window if the data input changes either from 1 to 0 or vice versa, then the flip-flop output will be metastable which indicates the hold violation. For more details, refer Chaps. 12 and 15.

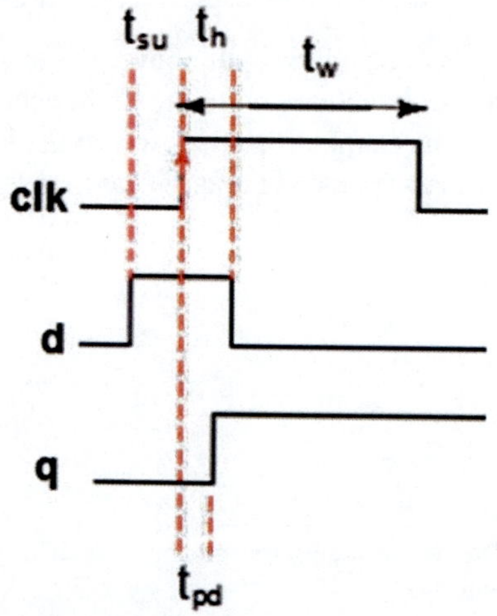

Fig. 5.1 Timing parameters for D flip-flop

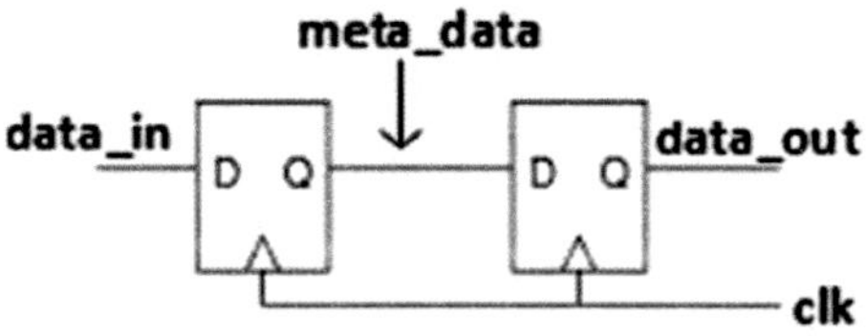

Fig. 5.2 Level synchronization concept

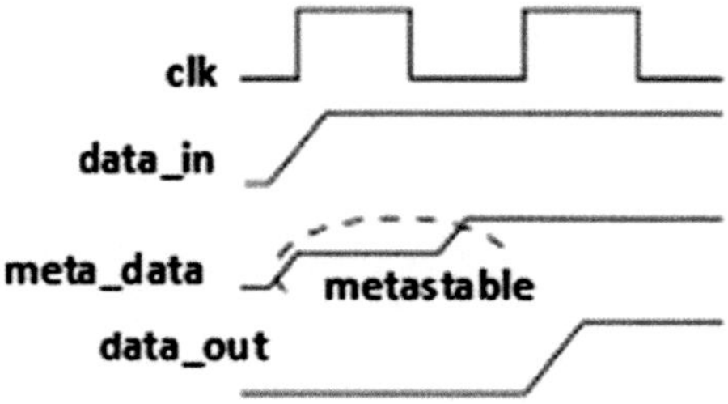

Fig. 5.3 Timing sequence for Fig. 5.2

- **Propagation Delay of Flip-Flop ($t_{pd} = t_{cq}$)**: The amount of time required for which the flip-flop to generate the valid output after the arrival of the active edge of the clock is called as propagation delay of flip-flop.
  The propagation delay is also called as the **clock to q delay**, and it is also referred as t_{cq}.

5.2 Metastability

If the data input of the design shown in Fig. 5.2 is connected to the other module whose clk is generated from the different clock sources, then the first flip-flop output will go into metastable state. The meta_data indicates the flip-flop data is metastable and hence there is timing violation for the first flip-flop.

The metastability indicates the data output is not valid and to get the valid data output the design needs to use the multiflop-level synchronizers.

The timing sequence of the data sampled by the first flip-flop and the output of the second flip-flop is shown in Fig. 5.3. As shown, the output of the first flip-flop is in the metastable state and the data_out output from the output flip-flop is having the valid legal state.

5.3 Clock Skew

If multiple clocks are in the ASIC design, then the clock distribution and clock tree synthesis will play very important role to balance the clock skew between the different clock inputs of various blocks.

Fig. 5.4 Synchronous design

The clock tree synthesis which is named as CTS in this book is discussed in Chap. 16. To get basic details about the clock network, let us understand about the clock skew.

If two different clocks in design arrive at the different time instances, then the design has the clock skew.

The reason for clock skew is the routing delay that is wire delay for the single clock domain design. Consider the figure shown, and let us consider that the clk edge at the launch flip-flop is arrived at time instance 't0' and at capture flip-flop at time instance 't2'. As the clock arrival time is different for this synchronous design, there is phase shift between the clk1 and clk2 and we can consider this as clock skew.

Another reason we consider is the aging of the oscillator; then, there is cycle-to-cycle frequency variation of the clock generated from the oscillator and hence difference in the arrival time can be named as jitter.

In Fig. 5.4, the clock skew is due to the interconnect delays between the clk1 pin and clk2 pin.

Practically in the ASIC design, we experience the two different types of skew (Fig. 5.5).

1. **Positive clock skew**: It indicates the launch flip-flop clock (clk1) is triggered first and then the capture flip-flop clock (clk2) arrives. Figure describes the t_{skew}, and it is the difference between the arrival times of clk1 and clk2. In other words, we can imagine the positive clock skew is the data and clock running in the same direction and positive clock skew is better for the setup time but not good for hold time as there is positive margin to manage for!
2. **Negative clock skew**: It indicates the launch flip-flop clock (clk1) is triggered last and the capture flip-flop clock (clk2) is triggered first. Figure describes the t_{skew}, and it is the difference between the arrival times of clk1 and clk2. In other words, we can imagine the negative clock skew is the data and clock running in the opposite direction and negative clock skew is better for the hold time but not good for setup time!

In the ASIC design we always experience the clock skew due to jitter or due to interconnect that is wire delays and the following are important points which we should know.

1. Positive clock skew is good for the setup time but bad for the hold time.

Fig. 5.5 Skew in the design

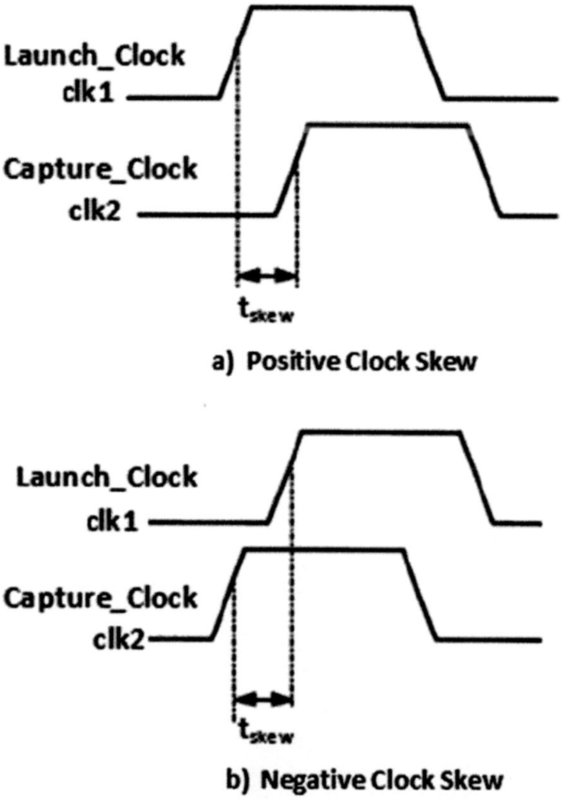

2. Negative clock skew is good for the hold time and bad for setup time.

5.3.1 *Positive Clock Skew*

As discussed before the positive clock skew where the launch flip-flop is triggered first and then capture flip-flop. There is margin of the buffer delay between the launch clock and capture clock and can be used to improve upon the required frequency for the design. The frequency calculation is discussed in Chap. 6.

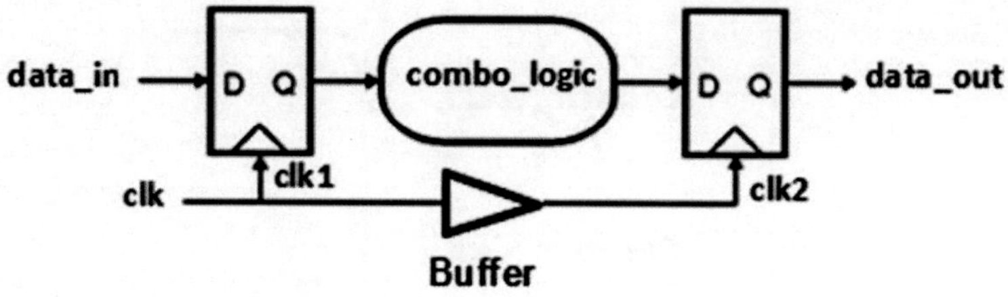

Fig. 5.6 Positive clock skew

Figure 5.6 shows the synchronous design with the positive clock skew and skew between the clk1 and clk2 is t_{buffer}.

Let us find the data required time and data arrival time.

$$\textbf{Data Arrival Time (AT)} = t_{\text{pff1}} + t_{\text{combo}}$$
$$\textbf{Data Required Time (RT)} = T_{\text{clk}} + t_{\text{buffer}} - t_{\text{su}}$$

where the T_{clk} is clock time period or clock to q delay, t_{buffer} is buffer delay, t_{su} is setup time of flip-flop, t_{pff1} is flip-flop propagation delay, and t_{combo} is combinational delay.

Setup slack is the difference between the data required time and data arrival time and should be positive. The positive setup slack indicates there is no any setup violation in the design.

To avoid the setup violations in the design, the design should have the fast data, fast launch clock (clk1), and slow capture clock (clk2). That is, the actual arrival of data should be fast as compared to the data required time (Fig. 5.7).

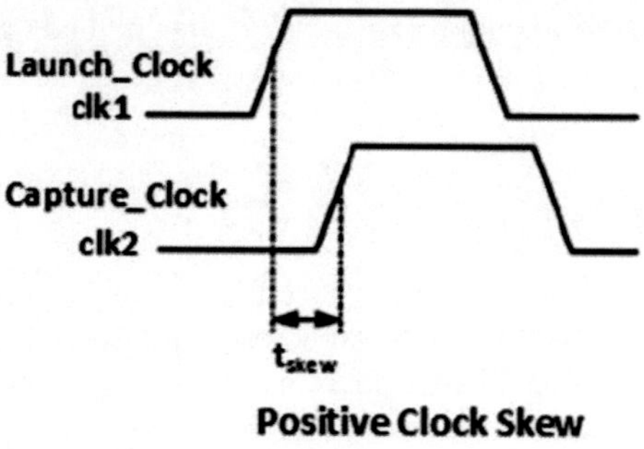

Fig. 5.7 Relationship between the launch and capture clock for the positive clock skew

5.3.2 Negative Clock Skew

As discussed before, the negative clock skew where the launch flip-flop is triggered last and the capture flip-flop is triggered first. As there is margin of the buffer delay between the launch clock and capture clock, this reduces the maximum frequency for the design. The frequency calculation is discussed in Chap. 6.

Figure 5.8 shows the synchronous design with the negative clock skew and skew between the clk1 and clk2 is t_{buffer}.

Let us find the data required time and data arrival time.

$$\textbf{Data Arrival Time (AT)} = t_{\text{buffer}} + t_{\text{pff1}} + t_{\text{combo}}$$
$$\textbf{Data Required Time (RT)} = T_{\text{clk}} - t_{\text{su}}$$

where the T_{clk} is clock time period or clock to q delay, t_{buffer} is buffer delay, t_{su} is setup time of flip-flop, t_{pff1} is flip-flop propagation delay, and t_{combo} is combinational delay (Fig. 5.9).

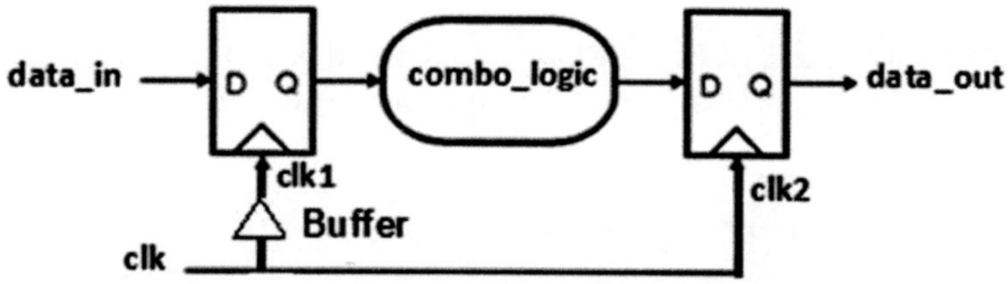

Fig. 5.8 Negative clock skew

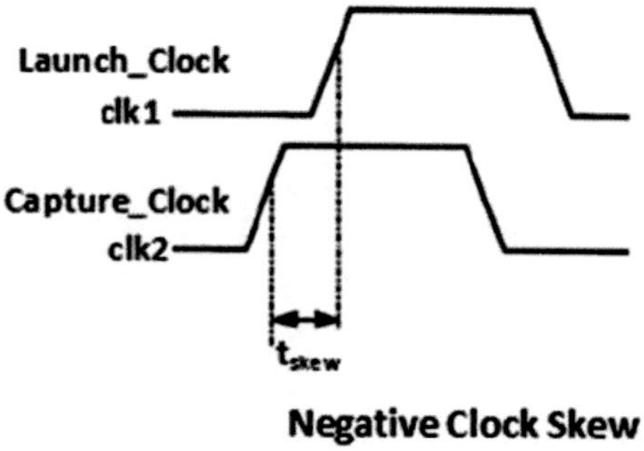

Fig. 5.9 Relationship between the launch and capture clock for the negative clock skew

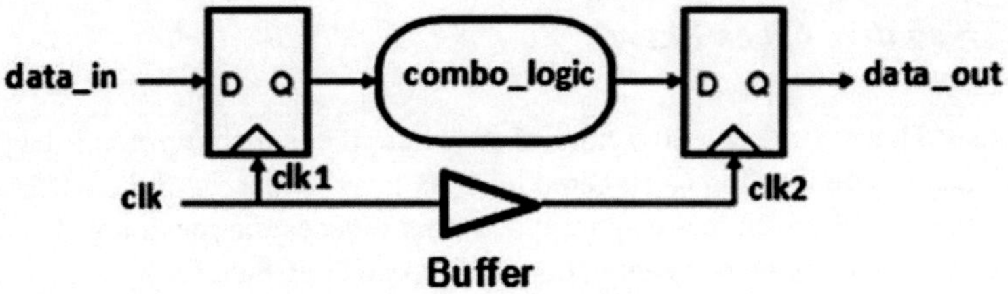

Fig. 5.10 Register-to-register path in synchronous design

5.4 Slack

During the ASIC design cycle, two terms are used for the slack that is setup slack and hold slack (Fig. 5.10).

5.4.1 Setup Slack

Setup slack is the difference between the data required time and data arrival time and should be positive. The positive setup slack indicates there is no any setup violation in the design.

$$\textbf{Data Arrival Time (AT)} = t_{\text{buffer}} + t_{\text{pff1}} + t_{\text{combo}}$$
$$\textbf{Data Required Time (RT)} = T_{\text{clk}} - t_{\text{su}}$$
$$\textbf{Setup Slack} = \textbf{RT} - \textbf{AT}$$

5.4.2 Hold Slack

Hold slack is the difference between the data arrival time and data required time and should be positive. The positive hold slack indicates there are no any hold violations in the design.

5.5 Clock Latency

The clock is generated form the PLL for the single clock domain designs and for the multiple clock domains we may need to have the multiple PLLs. The discussion on the PLL is ruled out during this book as it is analog component and needs to be designed by analog design team.

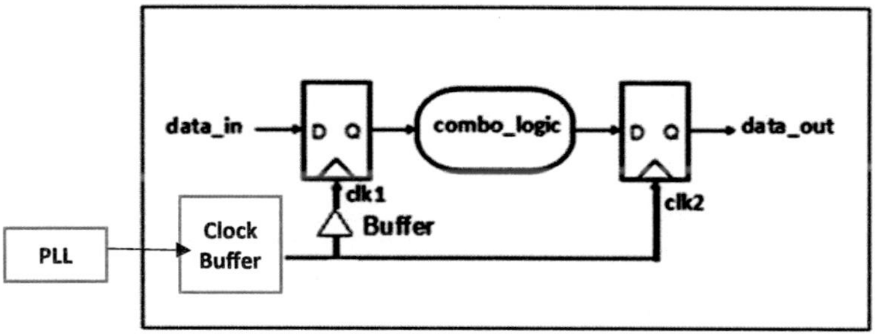

Fig. 5.11 Clock network latency

The clock network introduces the latency and it is effectively the time required for the clock to reach to the chip and the clock latency is due to the clock network delay during the clock distribution (Fig. 5.11).

5.6 Area for the Design

The overall area for the ASIC is due to the standard cell, macros, and IP cores. During the ASIC design of million or billion gates, the area constraints and better floor planning to get the desired performance play an important role. The area optimization we can think at various design phases such as:

1. During the architecture design by describing the better strategies for different functional block interactions
2. During the RTL design by using tool-based directives and commands and using the resource sharing techniques
3. In the physical design during the floor plan stage by having strategies to place the functional blocks to minimize routing delays and area due to use of the routing resources.

5.7 Speed Requirements

The speed is another important consideration during the design of ASIC. The ASIC performance can be improved by using different speed improvement techniques. For example, consider that the processor design works at the operating frequency of 500 MHz and we have the challenge to improve the design frequency. In such circumstances, various strategies can be used during the ASIC design cycle and few of them may be

1. Having better partitioning at the sequential boundaries during the architecture and micro-architecture design.
2. Having the initial floor plan where the interdependent blocks can be placed close to each other to minimize the area and hence the routing delay and to improve the speed.
3. During the RTL design stage, use the balance register and register duplication, optimization commands to improve on the design performance. But they may affect the logic area.
4. During the RTL design, use the registered inputs and outputs to have the better performance for the design.
5. Wherever it is feasible, use the pipelining concepts and architectures.
6. If FSM designs and controllers need to be used in the design, then try to work on the control and data path synthesis for the clean timing and better performance.
7. Try to use the synchronous designs as they are faster as compared to the asynchronous designs.
8. Try to avoid the internal clock generators; instead of that, think about the clock tree and optimize the clock tree during the CTS.
9. During the routing stage, try to work on the tool-based improvement techniques as enabling tool directives can play especially important role to balance the skews.

5.8 Power Requirements

For any kind of ASIC or SOC design, the important consideration is power and the objective of the design team is to reduce the leakage and dynamic power. The power planning by considering the power constraints is done during the physical design. The power optimization techniques should be used at the different stages during the ASIC design flow.

1. Having the low-power-aware architecture for the ASIC.
2. Use the UPF at various design levels.
3. During the RTL to minimize the dynamic power, use the dedicated clock gating cells.
4. The power can be also optimized during the RTL stage by avoiding unnecessary assignments and toggling of the data values.
5. Have the better power planning and power sequencing for the multiple power domains during the physical design.
6. Have the better strategies for the power shutdown during the physical design.

For more details, refer to Chap. 7.

5.9 What Are Design Constraints?

The design constraints are basically the design rule constraints and optimization constraints. The constraints we can consider as block-level constraints, top-level constraints, and chip-level constraints.

Design Rule Constraints (DRC): These constraints we can consider as the foundry-laid rules and should be met. During the physical design, we will carry out DRC to check for whether all the foundry-laid rules meeting or not. Layout is clean indicates no DRC violations. These constraints are mainly

- Transition
- Fanout
- Capacitance.

Optimization Constraints: These constraints are used during the design and optimization phases. These constraints are mainly

- Area
- Speed
- Power.

Mainly using the Synopsys DC, we will use the area and speed constraints and will try to optimize design during the various optimization phases.

The physical design tool such as Synopsys IC Compiler uses the area, speed, power constraints to meet the constraints to have the clean chip layout. These constraints are discussed in Chap. 10.

5.10 Exercises

Few of the exercises are useful during the architecture and RTL design of the processor.

1. Design the ALU to perform the arithmetic and logic operations on the two 8-bit numbers. Consider arithmetic operations as the addition, subtraction, increment, decrement, and logic operations such as OR, AND, XOR, and complement. Sketch the logic and try to work on the area optimization strategies using the digital design concepts.
2. Find the data required time and data arrival time for the following design (Fig. 5.12)?

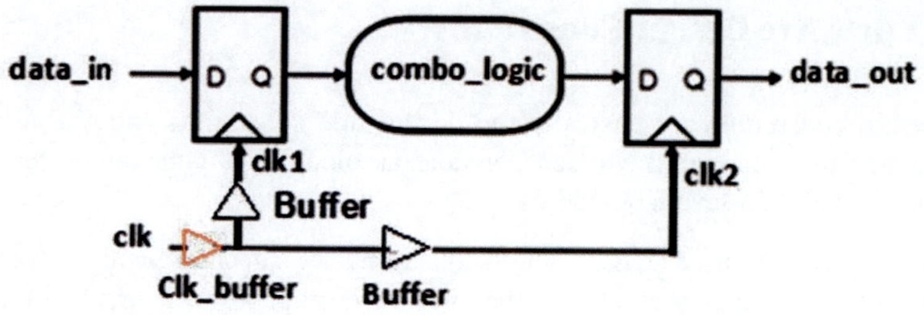

Fig. 5.12 Exercise 2

5.11 Chapter Summary

The following are few of the important points to conclude the chapter.

1. The minimum amount of time for which the data input of the flip-flop should maintain the stable value prior to arrival of the active edge of the clock is called as setup time.
2. The minimum amount of time for which the data input of the flip-flop should maintain the stable value after the arrival of the active edge of the clock is called as hold time.
3. The propagation delay of flip-flop is also called as clock (clk) to output (q), that is, t_{clktoq} delay.
4. Skew is the difference between the arrival times of clock.
5. The positive clock skew is the data and clock running in the same direction, and positive clock skew is better for the setup time but not good for hold time!
6. The negative clock skew is the data and clock running in the opposite direction, and negative clock skew is better for the hold time but not good for setup time!
7. Setup slack is the difference between the data required time and data arrival time and should be positive.
8. Hold slack is the difference between the data arrival time and data required time and should be positive.

Chapter 6
Important Considerations for ASIC Designs

The ASIC design phase of the architecture and micro-architecture design plays the important role to yield better performance for the ASIC chip. The design should have the less area, more speed, and low power that are demand of the chip. In such scenarios, we need to understand the following considerations to finalize the architecture.

1. Clocking sources
2. Clock latency and network latency
3. Single and multiple clock domain designs
4. Low power-aware architecture
5. Impact of skew on the speed
6. Clocking and reset policies
7. Synchronization and data integrity checks.

Few of these concepts are discussed in the previous few chapters, and few will be discussed in the chapter so that readers can have better understanding of the architecture and micro-architecture designs!

6.1 Synchronous Design and Considerations

As discussed in Chaps. 4 and 5 we always used the synchronous design as asynchronous designs are prone to the glitches and slower. The asynchronous clocking introduces the delays in the triggering of the unit and not recommended in the ASIC designs. Practically to understand the synchronous design, let us try to understand the logic shown in the figure. As shown, the register to register (reg to reg) path frequency is one of the deciding factors for the synchronous design.

ASIC is Application-Specific Integrated Circuits. The understanding of the important design considerations plays an important role during design phase!

V. Taraate, *ASIC Design and Synthesis*,
https://doi.org/10.1007/978-981-33-4642-0_6

Fig. 6.1 Synchronous sequential circuit

We have the source flip-flop which launches the data, and the capture flip-flop captures the data. The launch clock is clk1, and capture clock is clk2. The maximum operating frequency for the design indicates that the frequency for which all the timing paths in the design doesn't have any kind of timing violations. The timing paths are discussed in the next session and during the constraint definition in Chap. 10.

What is requirement for correct functionality?

The destination flip-flop data should be present prior to arrival of the active edge of the clock the design will not have any kind of the setup violation and that is the requirement for the design. So, the data required time (RT) is $T - t_{su}$, and the actual data arrival time (AT) depends on the propagation delay of D flip-flop ($t_{ctoq} = t_{pff}$) and the combinational delay (t_{combo}). So, the actual data arrival time is ($t_{ctoq} + t_{combo}$), and this limits the overall frequency for the design.

What is maximum clock frequency for synchronous design?

The setup slack is difference between the RT and AT and should be positive to avoid the timing violations in the design.

$$\text{Setup Slack} >= 0$$
$$\text{RT} - \text{AT} >= 0$$
$$(T - t_{su}) - (t_{ctoq} + t_{combo}) >= 0$$
$$T = t_{ctoq} + t_{combo} + t_{su}$$

The flip-flop timing parameters are t_{su} as setup time, t_{ctoq} as flip-flop delay and t_{combo} as combinational delay.

where T is clock time duration and the f_{max} is maximum frequency for the design which is equal to $f_{max} = 1/T$ (Fig. 6.1).

6.2 Positive Clock Skew and Impact on Speed

As discussed in the previous chapters, the positive clock skew is better for the setup as the capture flip-flop is triggered after the buffer delay but not good for the hold.

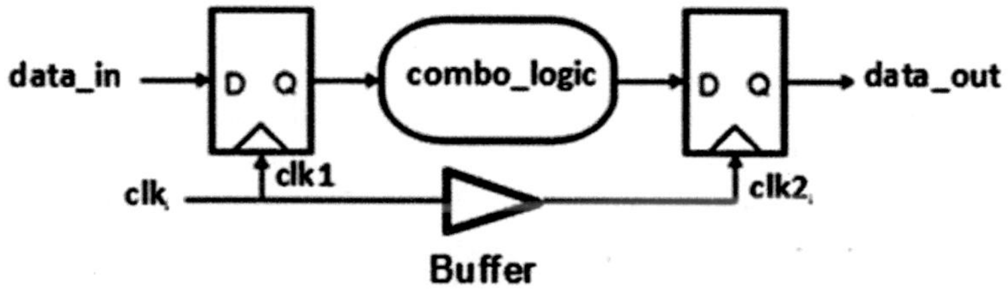

Fig. 6.2 Positive clock skew

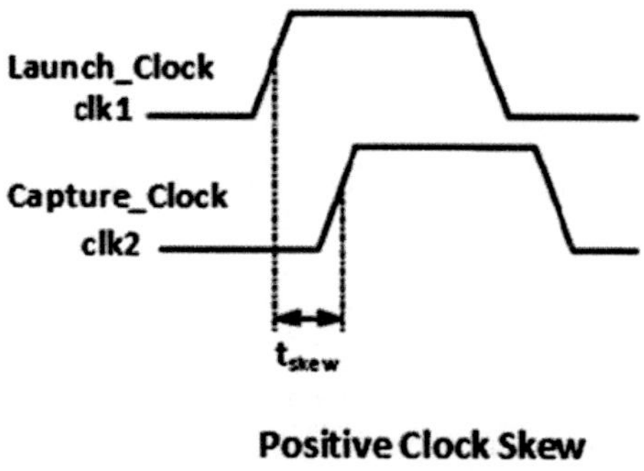

Fig. 6.3 Timing diagram for positive clock skew

The register to register path with the positive clock skew between the clk1 and clk2 is shown in Fig. 6.2. As shown, the clk2 is triggered after the t_{buffer} time delay and hence the chance of improving the clock frequency for the design (Fig. 6.3).

Consider the flip-flop timing parameters for the required technology node; the design works at the maximum clock frequency and the maximum frequency calculation is shown below.

$$\text{Setup Slack} >= 0$$
$$\text{RT} - \text{AT} >= 0$$
$$(T + t_{buffer} - t_{su}) - (t_{ctoq} + t_{combo}) >= 0$$
$$T = t_{ctoq} + t_{combo} + t_{su} - t_{buffer}$$

The flip-flop timing parameters are t_{su} as setup time, t_{ctoq} as flip-flop delay t_{combo} as combinational delay, t_{buffer} as buffer delay.

6.3 Negative Clock Skew and Impact on the Speed

In the negative clock skew the source flip-flop is triggered last and the capture flip-flop is triggered first. Here, we assume that the source flip-flop is the launch flip-flop and capture flip-flop as destination flip-flop.

The register to register path with the negative clock skew between the clk1 and clk2 is shown in Fig. 6.4. As shown, the clk1 is triggered after the t_{buffer} time delay, and hence, it reduces the overall operating frequency for the design (Fig. 6.5).

Consider the flip-flop timing parameters for the required technology node. The design works at the maximum clock frequency, and the maximum frequency calculation is shown below.

$$\text{Setup Slack} >= 0$$
$$\text{RT} - \text{AT} >= 0$$
$$(T - t_{\text{buffer}} - t_{\text{su}}) - (t_{\text{ctoq}} + t_{\text{combo}}) >= 0$$
$$T = t_{\text{ctoq}} + t_{\text{combo}} + t_{\text{su}} + t_{\text{buffer}}$$

The flip-flop timing parameters are t_{su} as setup time, t_{ctoq} as flip-flop delay and t_{combo} as combinational delay, t_{buffer} as buffer delay.

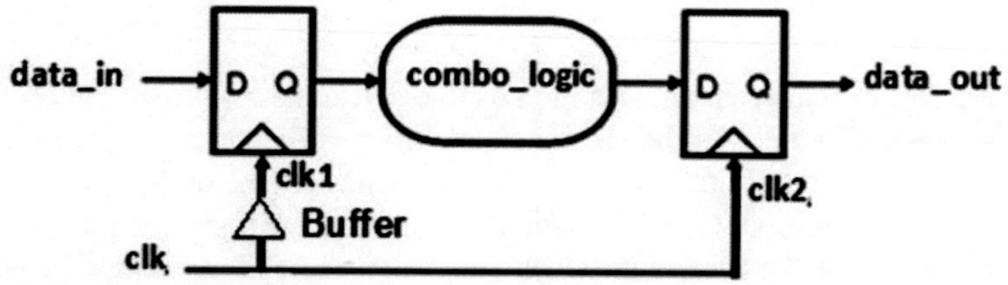

Fig. 6.4 Negative clock skew

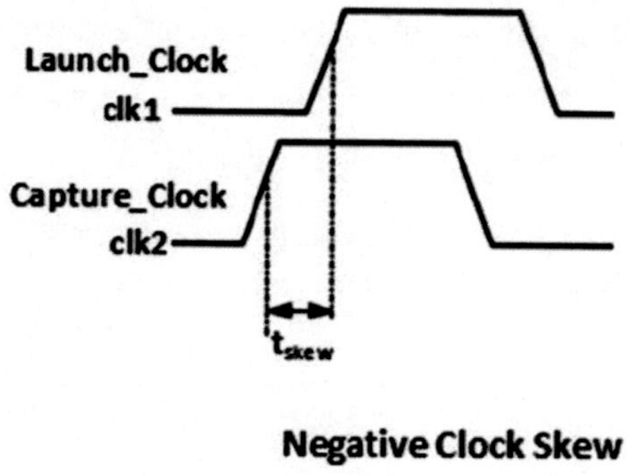

Fig. 6.5 Timing diagram for negative clock skew

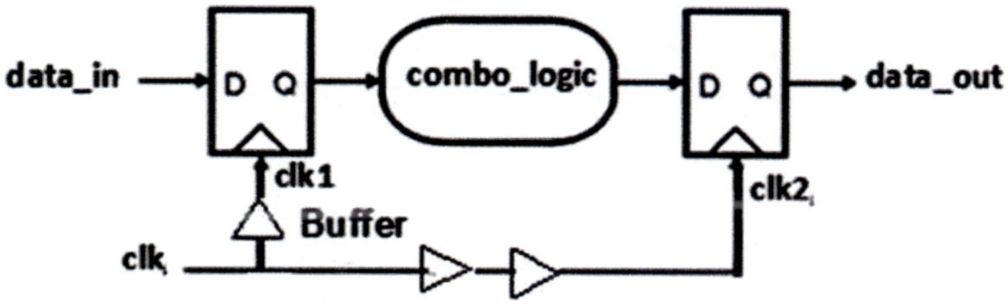

Fig. 6.6 Buffers in the clock network

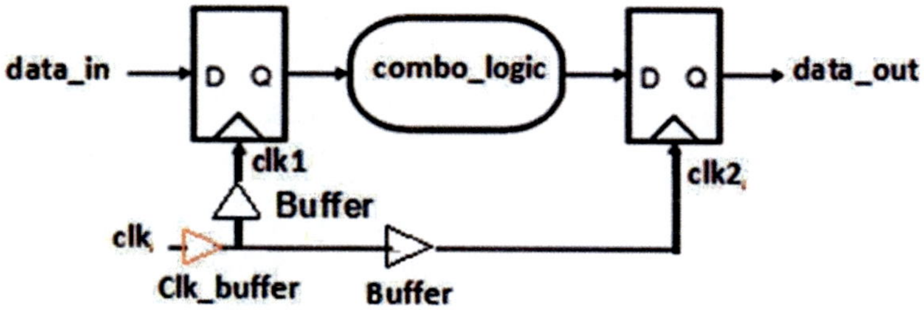

Fig. 6.7 Clock network latency and buffers

6.4 Clock and Network Latency

The clock network introduces the latencies in the design, and during the physical design, it is essential to optimize for the clock tree so that the distribution of clock can be with uniform clock skew.

If the clock tree strategies are not efficient and tool is not able to optimize for the clock network, then due to issues in the clock propagation, the design may have the issues due to timing violations.

It is real nightmare for the physical design team to understand the fix issues during the signoff static timing analysis (Figs. 6.6 and 6.7).

6.5 Timing Paths in the Design

Let us try to understand the timing paths in the synchronous design. Consider the design shown in Fig. 6.8 which has mainly four timing paths, and they are named as

1. Input to reg path
2. Reg to output path
3. Reg to reg path
4. Input to output path.

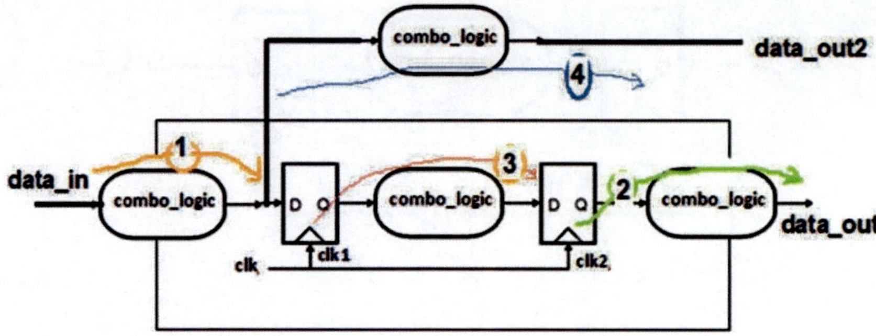

Fig. 6.8 Synchronous design

Fig. 6.9 Input to reg path

To identify the timing paths in the design, the designer should know the start point and end point.

Start point: The clock input of the sequential element (clk), data inputs of the sequential design is treated as the start point, and the tool algorithm identifies initially the start points for the design and then end points.

End Point: The end point of the design is the output of the sequential element or data input of the sequential element D flip-flop (D).

6.5.1 *Input to Reg Path*

It marked as path 1 in the figure, and it is from the input port data_in of the design to the D input of the sequential element (Fig. 6.9).

6.5.2 *Reg to Output Path*

It marked as path 2 in the figure, and it is from the clock pin clk2 of the flip-flop to the data_out1 of the sequential element (Fig. 6.10).

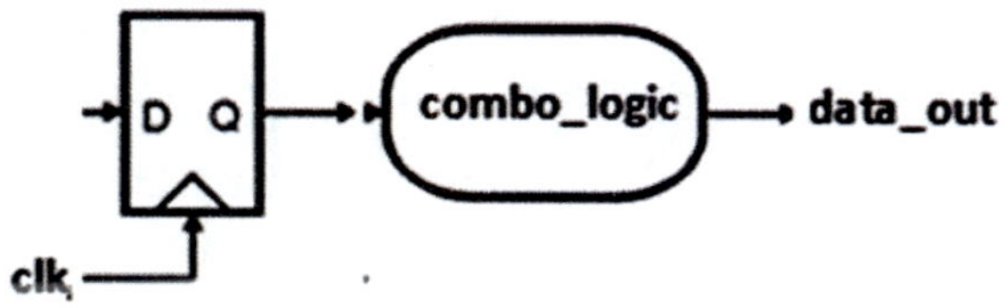

Fig. 6.10 Reg to output path

Fig. 6.11 Reg to reg path

Fig. 6.12 Combinational path

6.5.3 Reg to Reg Path

It marked as path 3 in the figure, and it is from the clock pin clk1 of the flip-flop to the data input of the D flip-flop that is sequential element 2 (Fig. 6.11).

6.5.4 Input to Output Path

It is unconstrained path and also called as combinational path. It is marked as path 4, and it is from data_in of the design to the data_out2 of the design (Fig. 6.12).

6.6 Frequency Calculations

To find the maximum operating frequency for the design Fig. 6.13, consider the timing parameters of flip-flop as

$$t_{\text{ctoq}} = 2\,\text{ns}$$
$$t_{\text{combo}} = 2\,\text{ns}$$

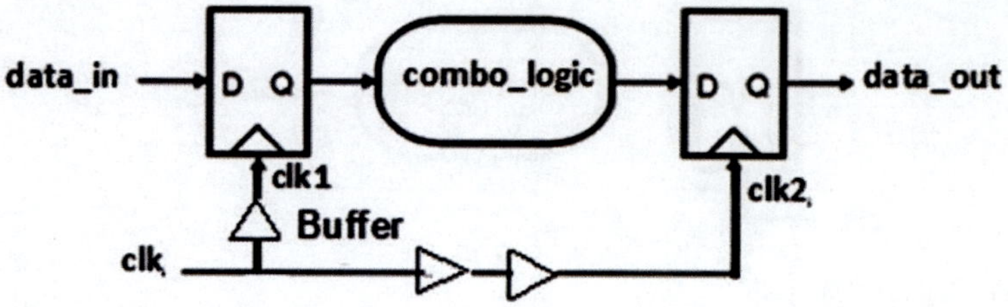

Fig. 6.13 Design example-1

$$t_{su} = 1\text{ ns}$$
$$t_{buffer} = 1\text{ ns}$$
$$t_h = 0.5\text{ ns}$$

For the design shown below, the data arrival time is equal to

$$\begin{aligned} AT &= t_{ctoq} + t_{combo} + t_{buffer} \\ &= 2\text{ ns} + 2\text{ ns} + 1\text{ ns} = 5\text{ ns} \end{aligned} \quad (6.1)$$

The data required time is (RT)

$$\begin{aligned} RT &= T - t_{su} + 2 * t_{buffer} \\ &= T - 1\text{ ns} + 2 * 1\text{ ns} \\ &= T + 1\text{ ns} \end{aligned} \quad (6.2)$$

Setup slack is RT-AT and should be greater than or equal to zero.

Using Eqs. 6.1 and 6.2, we get the T as 4 ns; therefore, the operating frequency for the design is 250 MHz.

For the design shown in Fig. 6.14, the clock buffer delay is common and hence can be considered as clock latency from the source and can be specified in the constraints so data arrival time is equal to

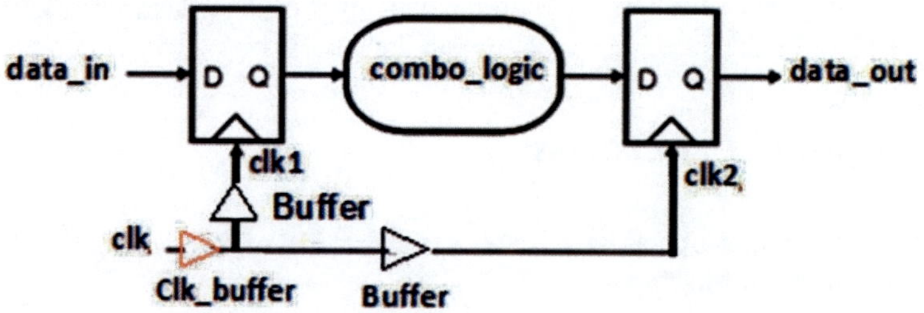

Fig. 6.14 Design example-2

$$\begin{aligned} AT &= t_{ctoq} + t_{combo} + t_{buffer} \\ &= 2\,\text{ns} + 2\,\text{ns} + 1\,\text{ns} \\ &= 5\,\text{ns} \end{aligned} \tag{6.3}$$

The data required time is (RT)

$$\begin{aligned} RT &= T - t_{su} + t_{buffer} \\ &= T - 1\,\text{ns} + 1\,\text{ns} \\ &= T \end{aligned} \tag{6.4}$$

Setup slack is RT-AT and should be greater than or equal to zero.

Using Eqs. 6.3 and 6.4, we get the T as 5 ns; therefore, the operating frequency for the design is 200 MHz.

6.7 What Is On-Chip Variations

Practical ASIC designs will have the on-chip variation that is also called as PVT variation where P is process, V is voltage and T is temperature.

- **Process**: This specifies the process, and in simple words, it indicates the oxide thickness, device length and other parameters as the doping concentrations.
- **Voltage:** Every chip operates at specific voltage, and this parameter indicates the voltage level at which chip operates.
- **Temperature:** This parameter indicates the operating temperature for the chip.

The variation of the operating parameters PVT is responsible to have the different operating conditions such as

1. Min
2. Max.

That is, in other words, we need to work on the best case, worst case and typical case scenarios during design synthesis and timing analysis.

The operating conditions are defined in the library, and they indicate the PVT and delays. The library is characterized for the specific operating conditions. For the on-chip variation that is for another operating condition, the derate factor can be used during the synthesis and timing analysis.

Library design and cell characterization is out of scope as per objective of this book is concern, but the library developers can specify the number of operating conditions in the library.

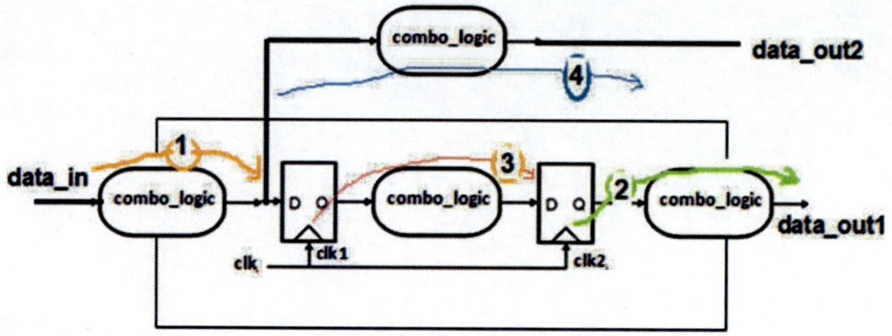

Fig. 6.15 Exercise-1

6.8 Exercises

1. Consider the design shown in the figure and find out maximum frequency for the design. Consider timing parameters as? (Fig. 6.15)

$$t_{ctoq} = 2\,\text{ns}$$
$$t_{combo} = 2\,\text{ns}$$
$$t_{su} = 1\,\text{ns}$$
$$t_{buffer} = 1\,\text{ns}$$
$$t_h = 0.5\,\text{ns}$$

6.9 Chapter Summary

The following are important points to conclude the chapter.

1. Use the synchronous design as asynchronous designs are prone to the glitches and slower.
2. In the negative clock skew, the source flip-flop is triggered last and the capture flip-flop is triggered first.
3. In the positive clock skew, the source flip-flop is triggered first and the capture flip-flop is triggered last.
4. Four timing paths are named as

- Input to reg path
- Reg to output path
- Reg to reg path
- Input to output path

5. ASIC designs will have the on-chip variation that is also called as PVT variation where P is process, V is voltage and T is temperature.

Chapter 7
Multiple Clock Domain Designs

Consider the ASIC design scenario in which the requirement is to have the different blocks for the complex designs, and few of these blocks are

1. Processor
2. Memories
3. Floating-point engine
4. Memory controllers
5. Bus interfaces
6. High-speed interfaces.

Consider that the processor and memories working on the operating frequency of 500 MHz and the floating-point engine with the memory controller works at the operating frequency of the 666.66 MHz. The bus interfaces and high-speed interfaces working on the operating frequency of 250 MHz that is the design have the multiple clocks and are treated as multiple clock domain design.

7.1 General Strategies for Multiple Clock Domain Designs

As discussed earlier, the issue in the multiple clock domains is while transferring of the data and control signals, and it has impact on the data integrity. The following strategies can be helpful during the ASIC design phase.

1. Try to have strategies for the data and control path optimization.
2. Try to have the strategy to define are create the multiple clock domain groups.
3. Try to have strategies to deploy the synchronizers while passing the control signals between the multiple clock domains.

ASIC chips have many clocks, and clock domain managament is important during the ASIC design cycle.

V. Taraate, *ASIC Design and Synthesis*,
https://doi.org/10.1007/978-981-33-4642-0_7

4. Try to use the data path synchronizers using FIFO and buffers to improve the data integrity.

The subsequent sessions discuss important issues and the strategies and their use during the multiple clock domain designs.

7.2 What Are Issues in the Multiple Clock Domain Design

If we consider the moderate gate count design or the processor core which works using the single clock and the design, it may have the timing violations during the layout stage due to additional interconnect delays. But such kind of designs may meet the timing and performance through the architecture, RTL, synthesis, and the tool-based optimization tweaks.

Now consider the design which demands the need of the multiple clocks as shown in the Figure and let us try to understand the issues in the design!

1. Due to multiple clock domains, the data integrity is major issues and design needs to go through the data integrity checks.
2. The flip-flops at the boundary of the clock domains without the use of synchronizers will have the metastability issues due to setup and hold violations.
3. The design will have timing violations, and it becomes difficult to force the sequential circuit output into valid legal states.

Let us understand the above using the sequential circuit which has multiple clock domains. Due to phase difference between the arrival of the clk1 and clk2, the flip-flop in the second clock domain will have the setup and hold violations that is the flip-flop output data_out will be metastable. The reason being the q output from the clock domain 1 can change during the setup and hold window of the clk2 active edge, and hence the data_out will be forced into the illegal state that is a metastable state (Fig. 7.1).

The timing sequence is shown in Fig. 7.2.

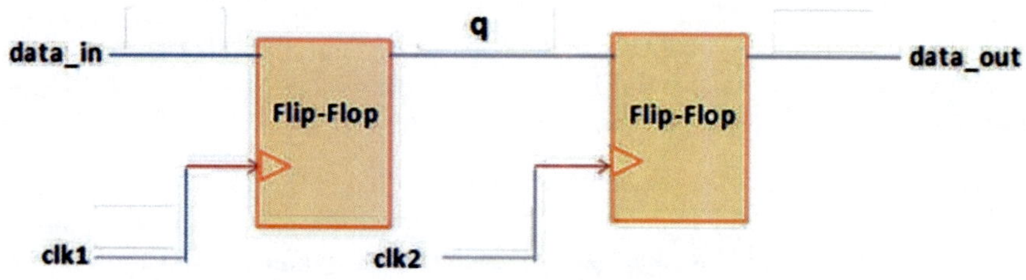

Fig. 7.1 Multiple clock domain concept

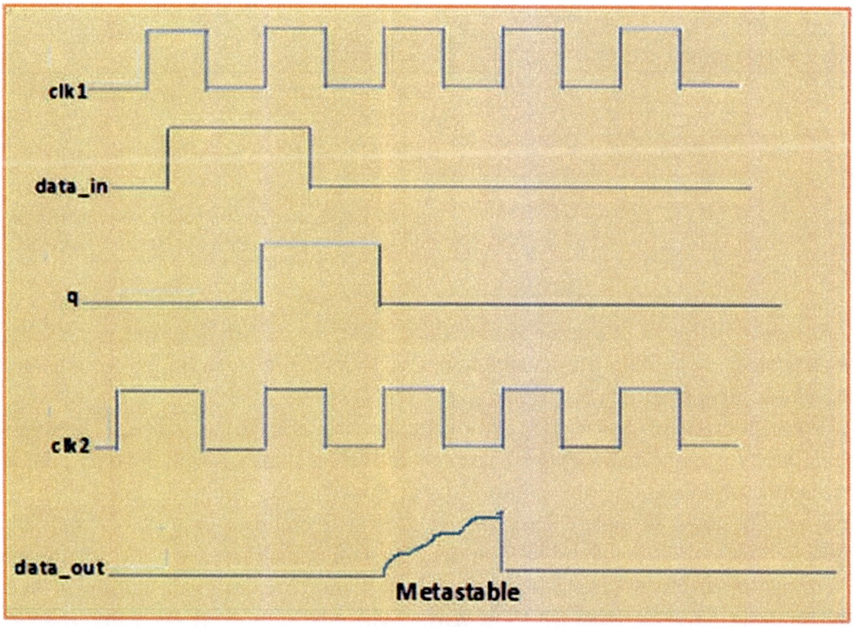

Fig. 7.2 Metastable output

7.3 Architecture Design Strategies

Consider the design which has three clock domains as shown in Fig. 7.3, and Table 7.1 describes information about the clock domains operating at various clock frequencies. Refer Chap. 9 for more information on the architecture and micro-architecture design.

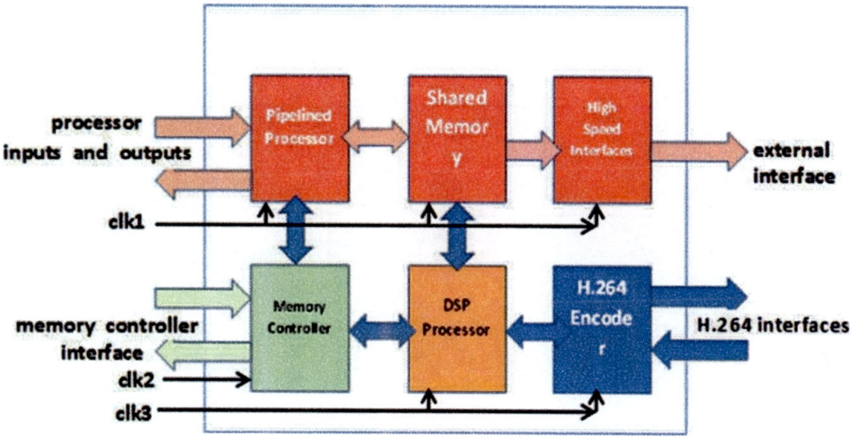

Fig. 7.3 Multiple clock domain architecture

Table 7.1 Multiple clock domain clock groups

Clock domain control	Frequency in MHz	Description
clk1	500	The clock domain one operating at frequency of 500 MHz
clk2	666.66	The clock domain two operating at frequency of 666.66 MHz
clk3	250	The clock domain three operating at frequency of 250 MHz

As a designer or architect, we need to think about the overall data integrity checks for the multiple clock domain design and need to have the clean timing for the data path and control path.

By considering this, we need to deploy the synchronizers to carry the data between the multiple clock domain designs. The synchronizers such as level, mux, and pulse are useful while passing the control signal between the multiple clock domain designs. The asynchronous FIFO can be used as synchronizer to carry the data between the clock domain and used in the data path.

Following are few of the guidelines which we should use during the multiple clock domain designs to eliminate the CDC errors, where CDC is clock domain crossing

1. **Avoid Metastability**: While passing the control signal information, use the registered output as this is useful to avoid the glitches and hazards. The multiple transitions during single clock cycle can be avoided by using the registered output logic while passing the control signal. Metastability blocking logic is shown in Fig. 7.4.
2. **Use of MCP**: Multicycle path formulation is highly recommended to avoid the metastability issue while passing the data and control signal information between the multiple clock domains. In the MCP, the strategy is to create the control and data pairs to pass the multibit data with the single bit control signal from sending clock domain to receiving clock domain. The control information can

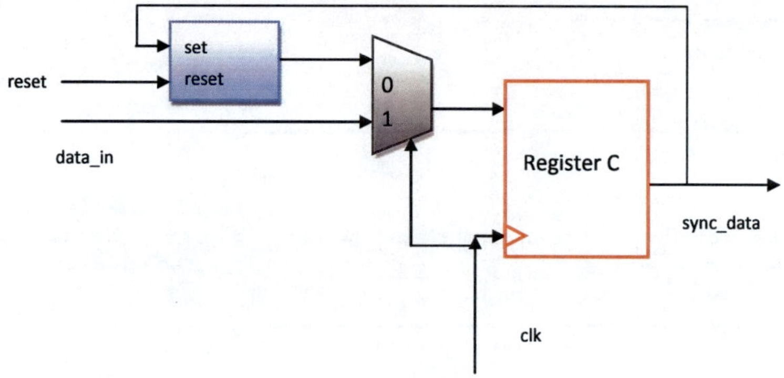

Fig. 7.4 Metastability blocking logic

be sampled in the receiving clock domain by using the pulse synchronizer, and data can be passed to the receiving clock domain with or without synchronizers. This technique is highly effective as the data can maintain the stable value for multiple cycles and can be sampled in the receiving clock domain by using the synchronized signal generated by using pulse synchronizer. Across the clock domain crossing boundaries, following are key points need to be considered.

(a) Control signals must be synchronized using the multistage synchronizers.
(b) Control signals should be free of hazards and glitches.
(c) There should be single transition across clock boundaries.
(d) Control signal should be stable for at least single clock cycle.
The MCP formulation is shown in Fig. 7.5.

3. **Use FIFO**: The effective technique to pass the multibit control signals or data information is using asynchronous FIFO. In this technique, the sending clock domain writes the data into FIFO memory buffer when the FIFO is not full and receiving clock domain reads the data from the FIFO buffer when the FIFO is not empty.

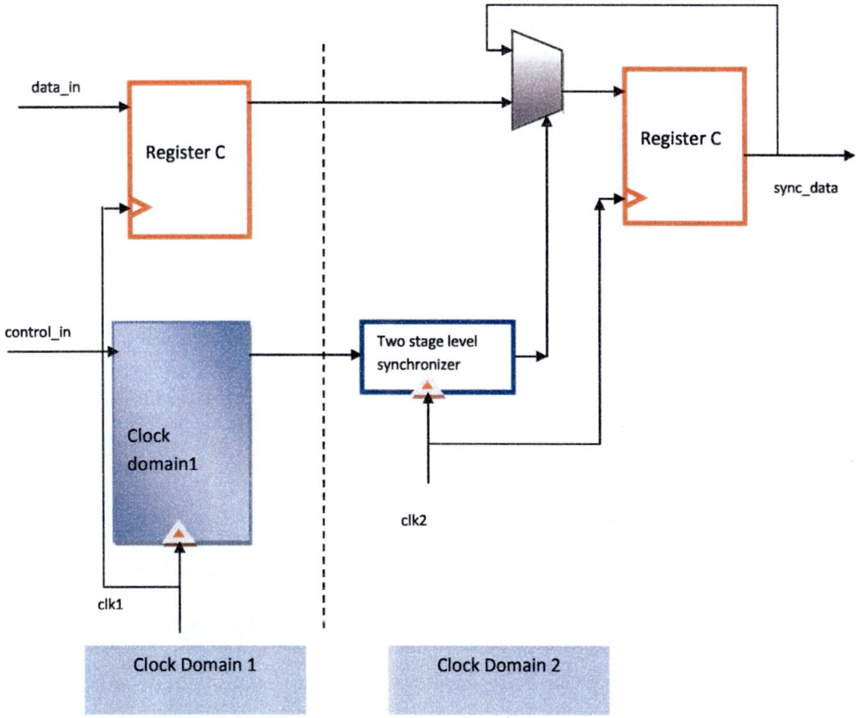

Fig. 7.5 MCP formulation

4. **Use Gray Code Counters**: In most of the ASIC designs with clock domain crossing (CDC), it is essential to pass the counter value across the clock domains. If binary counters are used to exchange the data at the clock domain boundaries, then due to toggling of one or more than one bit the data exchange can be error prone. In such scenarios, it is recommended to use the gray code counters to pass the data across the clock boundaries. In the receiver clock domain, use the gray to binary code converter to get the original data back!
5. **Design Partitioning**: While designing the logic for the multiple clock domain design, partition the design by using the clock groups.
6. **Clock Naming Conventions**: It is recommended to use the clock naming conventions to identify the clock source for better readability. The naming conventions for the clock should be supported by the meaningful prefix. For example, for sending clock domain, use clk_s, and for receiving clock domain, use clk_r.
7. **Reset Synchronization**: For the ASIC designs, it is highly recommended to use the reset synchronizers.
8. **Avoid Hold Time Violations**: To avoid the hold time violations, it is recommended to have close look at the architecture and have a strategy while passing the stable data between the multiple clock cycles.
9. **Avoid Loss of Correlation**: Across the clock domain boundary, there are several ways due to which loss of correlation can occur. Few of them are

   (a) Multiple bits on the bus
   (b) Multiple handshake signals
   (c) Unrelated signals

   To avoid this, use the clock intent verification technique as these techniques will ensure the passing of multibit signal across the clock boundaries.

7.4 Control Path and Synchronization

The section discusses various synchronizers and strategies to use them during the ASIC design.

7.4.1 Level or Multiflop Synchronizer

The control signals passing between the multiple clock domains mainly from the faster to slower clock domains will experience the timing failure, and the design will have timing violations. So, the better strategy during the architecture design is to identify the interface boundaries for the multiple clock domain design and then have the strategy in the RTL design to use the synchronizers.

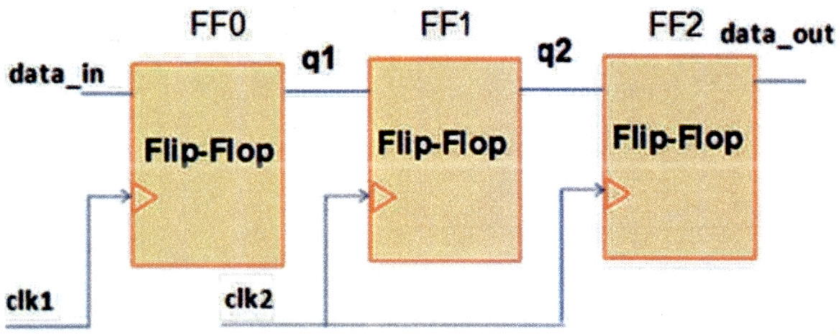

Fig. 7.6 Two-stage level synchronizer in the control path

The issue of metastability can be addressed by deploying the level synchronizers (may be y\using two or three flip-flops) while passing the control signals between the multiple clock domains. Figure 7.6 uses the two-stage level synchronizer logic.

As shown in Fig. 7.6, the level synchronizer is used to pass the control single q1 from clock domain 1 to clock domain 2. The main design strategy is to pass the valid output q1 to second clock domain. The level synchronizer is deployed in the second clock domain to sample the output q1. The input flip-flop in the second clock domain will be metastable due to violation of the setup or hold time, and this should be ignored by setting up the EDA tool attributes. The output data_out is valid data, and the design has the latency of two clocks due to use of the synchronizer.

The timing sequence for design shown in Fig. 7.6 is described below (Fig. 7.7).

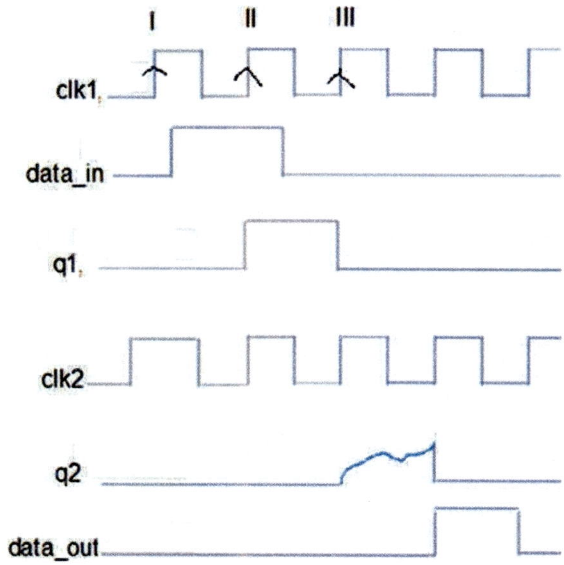

Fig. 7.7 Timing sequence with use of the two-stage synchronizer

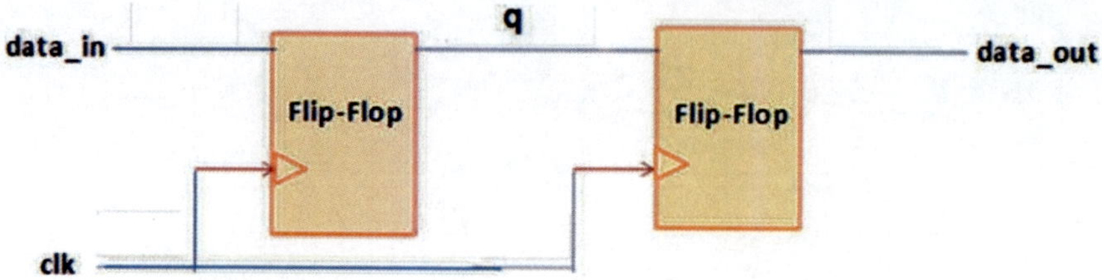

Fig. 7.8 Level synchronizer

As shown in Fig. 7.7, q1 is output from the first clock domain. On the rising edge of 'clk2,' the output of flip-flop FF1 that is q2 will go to the metastable state due to violation of either setup or hold time. But the flip-flop FF2 output that is data_out during the next subsequent clock edge is valid output. Set the false path using the command

```
set_false_path –from FF0/q –to FF1/q
```

The level synchronizers using the two flip-flops are shown in Fig. 7.8 and can be deployed in the design. The better strategy is to have the RTL description of the level synchronizers as separate module during the RTL design. The latency introduced depends upon the number of flip-flops required to drive the output into the valid legal state.

The piece of the RTL description is described below

```
always @ (posedge clk)
begin

        q<=data_in;
        data_out<=q

end
```

In the ASIC design, the issue of data integrity occurs when the control information needs to be passed from faster clock domain to the slower clock domain. The issue is due to non-converging of the legal states of the flip-flop outputs while passing the control signals from clock domain 1 to clock domain 2.

The issue of sampling the data from faster clock domain to the slower clock domain can be resolved by using pulse stretcher. The level to pulse generator which works on the positive clock edge is shown in Fig. 7.9.

Another mechanism which is handshake of signals can be used to have the data convergence.

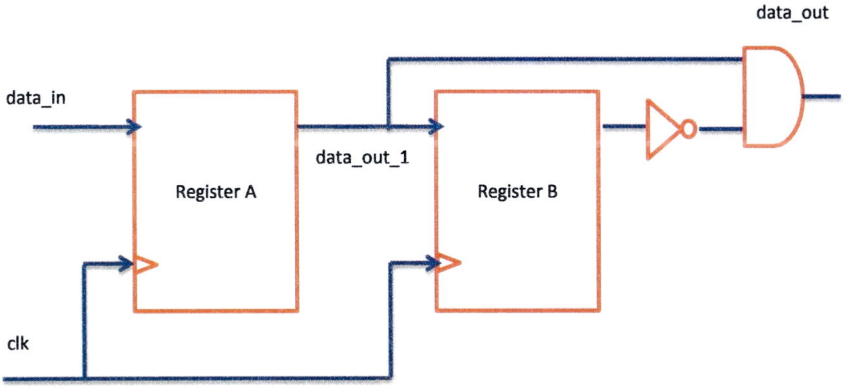

Fig. 7.9 Level to pulse conversion

As shown in Fig. 7.10, the sampled signal in the clock domain 2 is feedback as a handshaking signal to clock domain 1. This handshake mechanism is like acknowledgement or notification to the faster clock domain 1 that the control signal passed by the faster clock domain is successfully sampled by the slower clock domain. In most of the practical scenarios, this kind of mechanism is used, and even the faster clock domain can send another control signal after receiving the valid notification or acknowledgement signal from the slower clock domain.

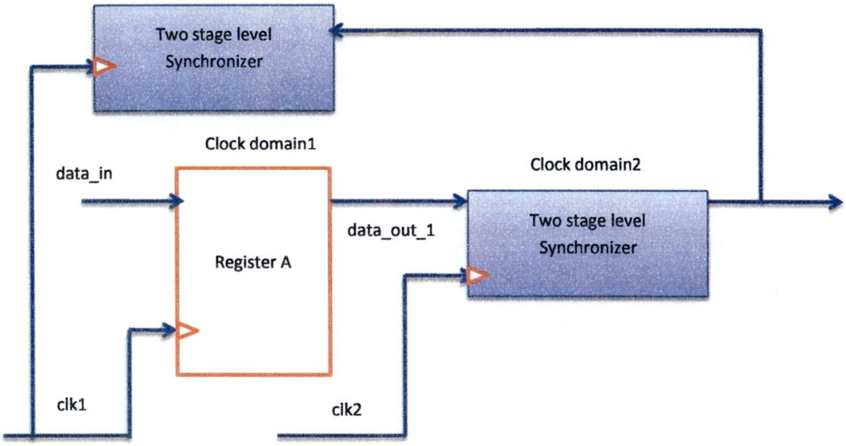

Fig. 7.10 Handshake mechanism for control signals

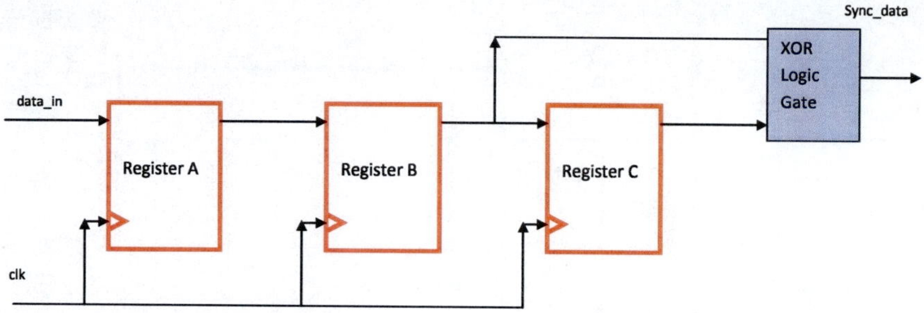

Fig. 7.11 Pulse synchronizer

7.4.2 Pulse Synchronizers

This type of synchronizer uses the multistage level synchronizer where the output of two-stage level synchronizer is sampled by the output flip-flop. This kind of synchronizer is also named as toggle synchronizer and used to synchronize the pulse generated in the sending clock domain into the destination clock domain. While passing the data from faster clock domain to the slower clock domain, the pulse can be skipped if two-stage level synchronizers are used. In such scenarios, the pulse synchronizers are efficient and useful. The pulse synchronizer diagram is shown in Fig. 7.11.

7.4.3 MUX Synchronizer

Use the pair of the data and control signals while sending the information from clock domain 1 to clock domain 2. Use the multiple bit data and use the single bit control signal. At the receiving end depending on the ratio of the sending clock and receiving clock, use the level or pulse synchronizer to generate the control signal for the multiplexer. This technique is like the MCP and effective if the data is stable for multiple clock cycles across the clock boundaries. The diagram is shown in Fig. 7.12.

7.5 Challenges in the Multiple Bit Data Transfer

Passing multiple control signals between the multiple clock domains is one of the important challenges. The issue is the different time of the arrival of these control signals If the arrival of these control signal is not managed properly, then the real issue is due to the skew. Consider the scenario shown in Fig. 7.13, where 'enable' 'load_en' and 'ready' need to be passed from one of the clock domains to another clock domain. In such scenario, if independent level synchronizers are used, then

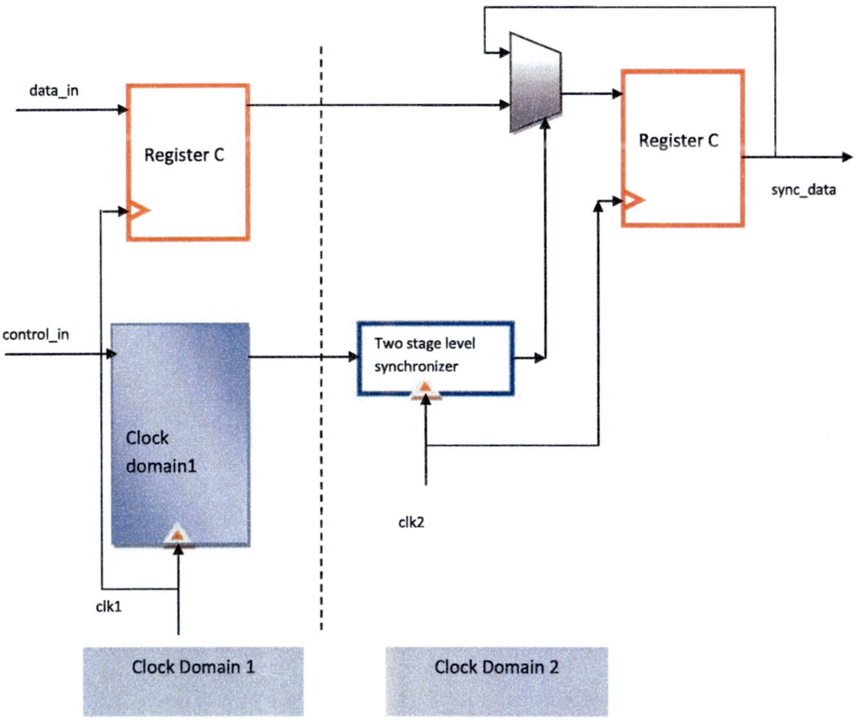

Fig. 7.12 Mux synchronizer

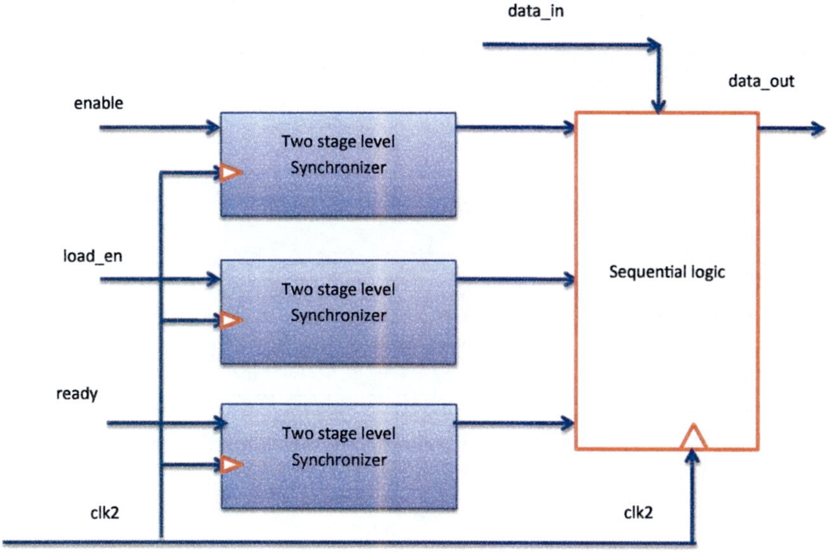

Fig. 7.13 Sampling of the multiple signals in the receiver clock domain

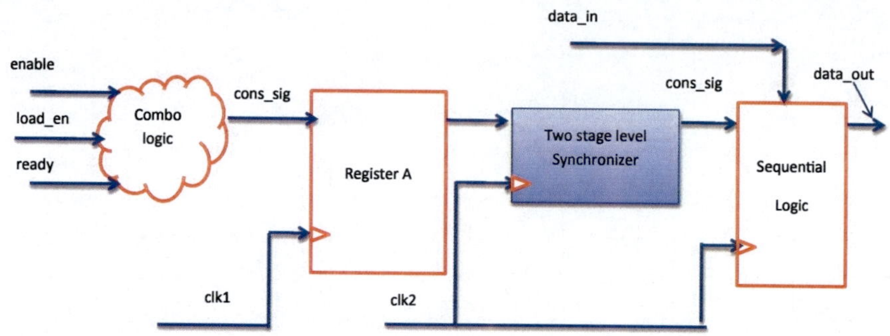

Fig. 7.14 Consolidated control signal passing in the multiple clock domain

there might be synchronization failure at the receiving end due to skew (different arrival time of these signals).

Consider that one of the control signal, for example, enable arrives late, then there may be synchronization failure in the control path, and to avoid this group these three control signals and try to pass the common signals between the clock domain. The strategy is shown in Fig. 7.14.

7.6 Data Path Synchronizers

The techniques used to pass the multiple bits of the data between the clock domains are

1. Handshaking mechanism
2. FIFO memory buffers.

7.6.1 Handshaking Mechanism

Use of the handshake mechanism is one of the techniques useful while passing of the multibit signals between the clock domain. Consider Fig. 7.15 as shown, the transmitter operates at clk1, and receiver operates at clk2. The data can be passed from transmitter to receiver.

Receiver clock domain can generate the handshake signals such as data valid and device ready. So, the purpose is to notify the transmitter that valid data is available on the bus and the device is not ready to receive the new data.

Handshake Signal Datavalid It is active high handshake signal from clock domain 2 and indicates that the data transmitted is valid data and receiver needs few clocks to sample this data. The clock latency while transferring the data is dependent upon

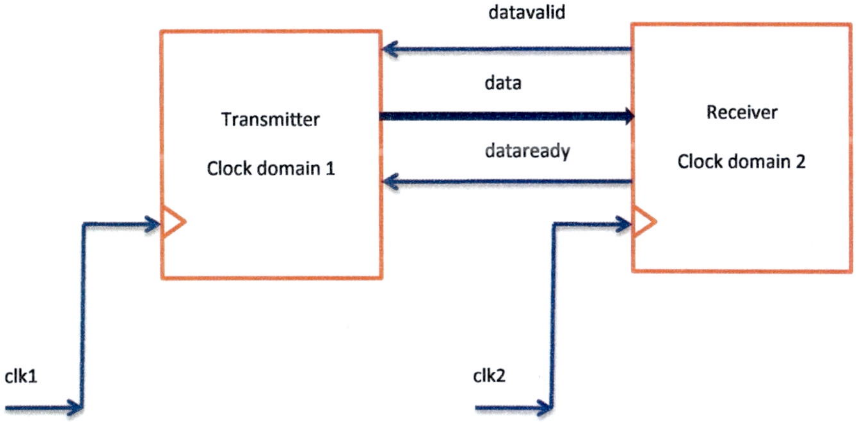

Fig. 7.15 Block diagram for handshake mechanism

the number of flip-flops used in the synchronizer, and the poor latency is one of the biggest disadvantages of the handshake mechanism.

Handshake Signal Deviceready It indicates that the receiver is ready to accept the new data when the data valid is de-asserted and the device ready can go high to notify transmitter that place new data on the data bus.

If we have the FSM controllers in the multiple clock domains, then design the architecture to establish synchronization by using the request and acknowledge (ack) signals. For the FSM control, handshake mechanism is shown in Fig. 7.16.

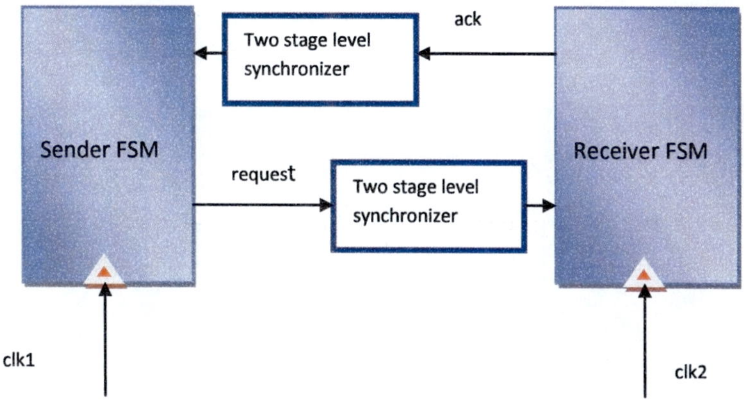

Fig. 7.16 FSM handshaking mechanism

7.6.2 *FIFO Synchronizer*

FIFOs are useful as data path synchronizer is used to exchange the data between multiple clock domains. The sender clock domain or transmitter clock domain can write the data into the FIFO memory buffer using write_clk if FIFO memory buffer is not full and receiver clock domain can read the data by using the read_clk if FIFO memory buffer is not empty (Fig. 7.17).

The FIFO consists of the following blocks

1. **Memory**: memory buffer
2. **Write Clock Domain**: write domain logic which is working on write_clk
3. **Read clock Domain**: read domain logic which is working on read_clk
4. **Flag Logic**: empty and full flag generation logic.

The FIFO with the associated logic blocks is shown in Fig. 7.17.

How to get the depth of the FIFO?

Consider the write clock domain works at operating frequency of 250 MHz and read clock domain at 100 MHz and no latency then to transfer the burst of the 50 bytes use following calculations

1. *Write clock time: T1 = 1/250 MHz = 4 ns.*
2. *The Time Required to Write Burst of 50 Bytes of data = 4 ns*50 = 200 ns*
3. *Read Clock Time: T2 = 1/100 MHz = 10 ns*
4. *Number of Reads with 10 ns = 200 ns/10 ns = 20*
5. *Depth of FIFO = 50–20 = 30 Bytes.*

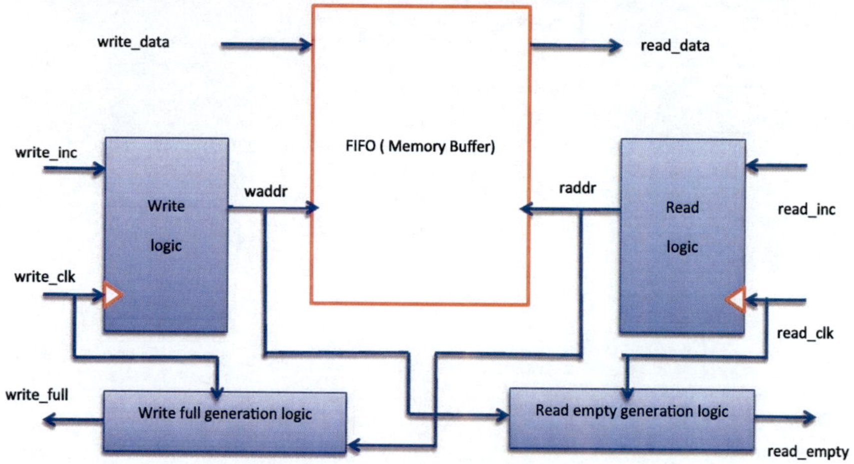

Fig. 7.17 Block diagram of FIFO

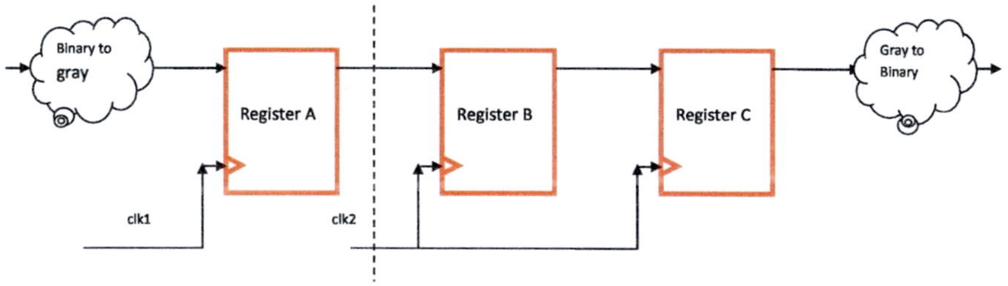

Fig. 7.18 Gray encoding technique

If read and write latency are specified, then try to modify the above steps to get the depth of FIFO.

7.6.3 *Gray Encoding*

While passing the multiple bit of the data or control signals, it is essential to use the gray encoding technique as this technique guarantees about the only one-bit change in two successive numbers.

For example, if 4-bit binary data needs to be passed between the multiple clock domains, then one or more than one bit toggles and hence more power is required and the chance of error. So to avoid this and to improve the performance, use the binary to gray code converter logic in the transmitter or sender clock domain. This guarantees only one-bit change across the clocking boundary. To get the original binary data in the receiver clock domain, use the gray to binary code converter. The technique is shown in Fig. 7.18.

7.7 Summary and Future Discussions

The following are few important points to conclude the chapters.

1. Deploy the data path synchronizers while passing the data between the multiple clock domains.
2. Deploy the control path synchronizers while passing the control signals between the multiple clock domains.
3. Multicycle path formulation is highly recommended to avoid the metastability issue while passing the data and control signal information across the clock domains.

4. The common and effective technique to pass the multibit control or data information is the use of asynchronous FIFOs.
5. For multibit control signal passing between the multiple clock domains, use the techniques by grouping these signals to avoid the skew due to different arrival time.

Chapter 8
Low Power Design Considerations

The ASIC design constraints are area, speed, and power. The issues related to the area and speed we have already discussed in the previous few chapters, and in the subsequent section we will discuss about the low-power designs. Following are few of the important goals of the ASIC designer to optimize for the power

1. Have the design strategies using the low-power cells.
2. Plan for the overall power requirements and have the better power planning during the physical design.
3. Try to optimize for the leakage and dynamic power using various techniques during power optimization.
4. Use the UPF at various stages with the power compilers.
5. Have the strategies in place to have the power sequencing and power shutdowns.

8.1 Introduction to Low Power Design

For the ASIC designs, the power optimization is especially important and overall power analysis and understanding play important role. To have the power optimization, the team works on the strategies to have the low-power design architecture. Power is basically dependent on the voltage, and during this decade the technology node has shrunk enough and hence the requirement of the core and IO voltage has reduced.

The power dissipation for the cell is $p = (1/2) * C_s * V^2 * f$. The power dissipation for any standard cell is directly proportional to the stray capacitance (C_s), applied voltage (V), and the frequency. To optimize for the power, we should have the lower voltage, lower capacitance, and low frequency. Practically, ASIC chips work at the higher frequencies, and reducing frequency is not the objective! There is always trade-off between the speed and power, and the architecture design should consider

Low-power aware designs and architecture are the real requirement during ASIC design.

V. Taraate, *ASIC Design and Synthesis*,
https://doi.org/10.1007/978-981-33-4642-0_8

the speed and frequency requirements to have the better architecture in place which can have the desired speed and the low power.

So, the power we consider is the leakage and dynamic power mainly. The primary source of power dissipation in CMOS is leakage current. The leakage current is summation of the cell leakage current and is state dependent.

$$P_{\text{leakage}} = \sum \text{Cell Leakage}$$

where cell leakage can be computed by using the library cell leakage, and it is state dependent.

The dynamic power is defined as addition of the summation of the cell dynamic power and summation of power dissipated due to wires. The following are the few equations which describes the leakage and dynamic power.

$$P_{\text{dynamic}} = \sum \text{Cell dynamic power} + \sum \frac{1}{2} * \text{Cl} * V * V * \text{Tr}$$

where the Cl is the capacitive load at pin or net, V is voltage level, and Tr is toggle rate.

Having the low-power design architecture is useful to have the low-power management and to have the long battery which is the real requirement. It is expected that the ASICs and devices should be of lightweight, small, cool, and even they should have the long battery life.

In simple words, let me describe this in the block diagram. We can imagine following to have the low-power aware architecture

1. Multiple power domains
2. Power sequencer and scheduling algorithms
3. Different cells such as
   (a) Level shifter
   (b) Isolation cells
   (c) Retention cells.

The basic architecture is described in Fig. 8.1. As shown, the architecture has the overall power management using the power sequencer/schedular and used to control the power domain I and power domain II.

8.2 Sources of Power

As discussed in the previous section, the power dissipation for the cell is $p = (1/2) * C_s * V^2 * f$. The power dissipation for any standard cell is directly proportional to the stray capacitance (C_s), applied voltage (V), and the frequency.

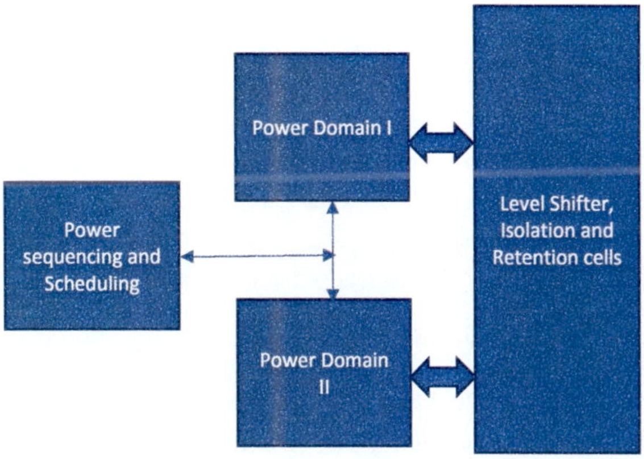

Fig. 8.1 Low-power design architecture

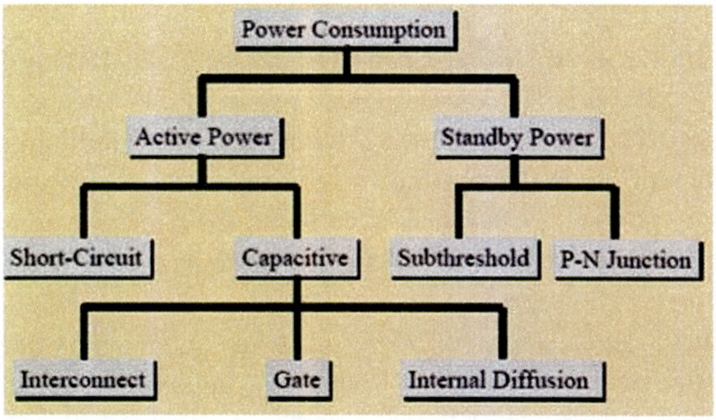

Fig. 8.2 Sources of power consumption

The sources of power consumption in CMOS are described in Fig. 8.2.

Following are few of the important points to understand the power dissipation in the CMOS

1. The power dissipation for any CMOS cell is function of the switching activity, capacitance, voltage, and the structure of transistor. So power is described as

$$\text{Power} = P_{\text{switching}} + P_{\text{short-circuit}} + P_{\text{leakage}}$$

2. The total power for any CMOS cell is summation of the dynamic and leakage power.
3. Dynamic power is summation of the switching power and short-circuit power.
4. The short-circuit power dissipation is due to the gate switching state, and it is due to the short circuit between the supply voltage and ground. The following equation describes the switching and short-circuit power

Table 8.1 Percentage power saving

Design abstraction stage	% Power saving
System design and architecture	70–80
Behavioral design	40–70
RTL design	25–40
Logic design	15–25
Physical design	10–15

$$P_{switching} = \alpha * f * C_{eff} * V_{dd} * V_{dd}$$

where α is switching activity, f is switching frequency, C_{eff} is effective capacitance, and V_{dd} is supply voltage.

$$P_{short\text{-}circuit} = I_{sc} * V_{dd} * f$$

where I_{sc} is short-circuit current during switching, f is switching frequency, and V_{dd} is supply voltage.

5. Dynamic power can be reduced by reducing the switching activity, clock frequency (it reduces the design performance), also by using the capacitance, and the supply voltage. If faster slew cells are used, then it consumes the less dynamic power and hence cell selection is important in reduction of the dynamic power.
6. Leakage power is given by the following equation, and it is function of the supply voltage V_{dd}, the switching threshold voltage V_{th}, and size of transistor.

$$P_{leakage} = f\left(V_{dd}, V_{th}, \frac{W}{L}\right)$$

In the above equation, the W is width of transistor and L is length of transistor.

Powers-saving opportunities at the different design phases are listed in Table 8.1.

8.3 Power Optimization During the RTL Design

As discussed earlier during the RTL design, the power can be optimized by 25–40% using various techniques and strategies. The section discusses the low-power design technique.

1. **Modeling and power estimation**: For the low-power design and the management of power for any SOC, it is essential to prepare the library models with the required power data. It is required to develop the transistor-level models for the custom blocks. The common practice in the SOC design at the RTL level is use of power

compiler to understand the power consumption based on the switching activity information from the RTL simulation data. This technique is useful for estimation of the power consumption at early stage of the design. Another important point to be considered at the gate level is to develop the glitch-free low-power designs and state and path dependencies support. As gate-level analysis is more accurate as compared to the RTL-level analysis, it is essential to use the time-based analysis based on the peak power and hot spots.

2. **Clock gating**: Use the clock gating technique using the clock gating cells to minimize the power during the RTL design. Clock gating can be implemented by identifying the synchronous load enable register banks. Clock gating can be implemented by using the gating of clock with the enables instead of recirculating of the data when enable is inactive. If power compiler is used at the RTL level, then it automatically optimizes the static, dynamic power dissipation with the delay and area to meet the design constraints.

Clock gating stops the clock and forces the original circuit in the zone of no transition. In the practical scenario, if we consider the functional block as

```
always@(posedge clk)
begin
if(enable)
data_out<=data_in;
end
```

The above piece of code generates the synthesis result shown in Fig. 8.3.

The above generated logic is without clock gating and has the higher-power dissipation. To reduce the power consumption, the clock gating logic needs to be used and can be designed by eliminating the multiplexers at the input, thus avoiding the recirculation of data. This results in the area and power savings and reduces the power consumption in the clock network. The synthesized logic using clock gating is shown in Fig. 8.4. The timing sequence is shown in Fig. 8.5.

The use of clock gating has drawback that the logic used to implement the clock gating technique is redundant and hence there can be issues in the testing and verification. Another important point need to keep in mind is that it is essential to stop the

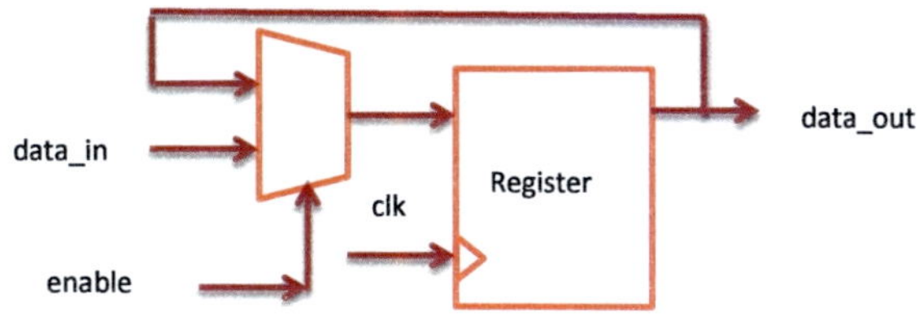

Fig. 8.3 Design without clock gating

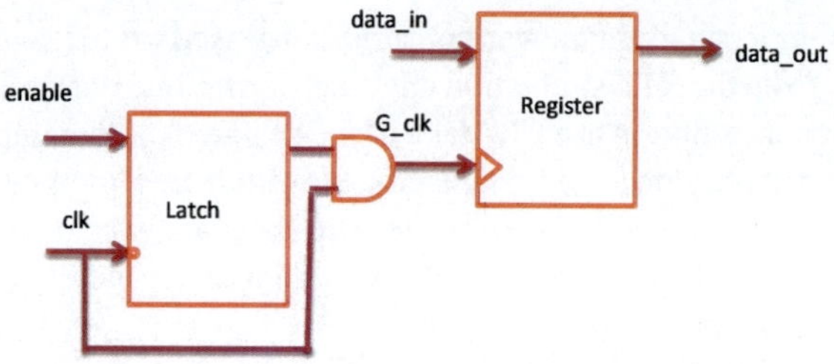

Fig. 8.4 Design with clock gating

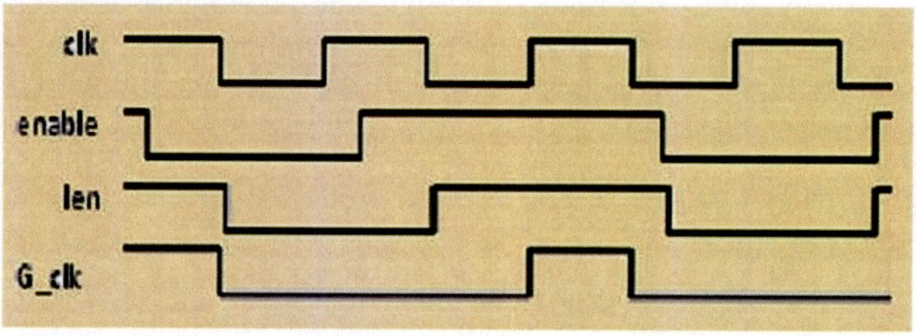

Fig. 8.5 Timing sequence for the clock gating

glitches and hazards on enable signal, and this is achieved by using the transparent latch between the enable and the AND logic gate.

Clock gating can be efficiently implemented by using the power compiler from Synopsys. Use the command ***set_clock_gating_signals***. Figure 8.6 illustrates the inputs and outputs used for the power compiler.

The outcome of the power compiler is the elaborated unmapped design. Power compiler uses the inputs as source RTL code and library to optimize for the low power.

The following are few of the key points need to be considered while implementing the clock gating using the power compiler.

1. General clock gating can be included or excluded from the design for the hierarchical modules. The command use is ***set_clock_gating_signals***. The care needs to be taken by the designer while using the power compiler for the same. Each design should have the single command line for both the inclusion and exclusion of the clock gating.
2. If the design has multiple registers and few of the registers need to be excluded from the clock gating strategy, then they should have the separate enable signal. If same enable signal is used, then it generates the same clock gating for the entire register bank. For example, if the data bus is defined as data_in[7:0] with the registered inputs and if the lower nibble data_in[3:0] need be excluded from clock gating, then it should have the different enable and data_in[7:4] should have different enable.

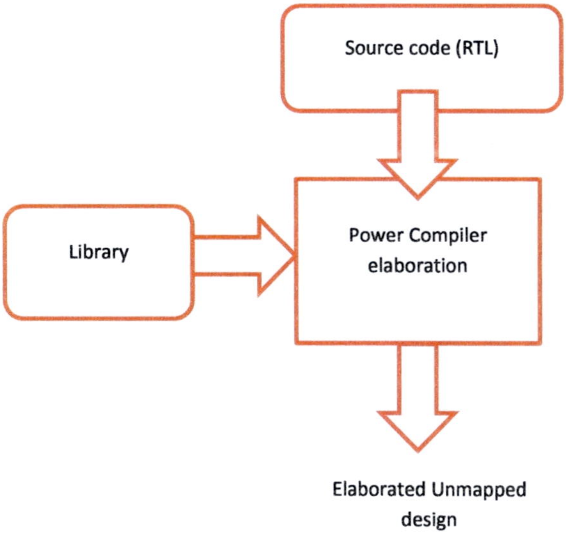

Fig. 8.6 Power compiler inputs and outputs

3. Clock gating signals as single bit or multiple bits have added advantage as it avoids the recirculation of the data by removing the multiplexers. But it can consume more area and additional power due to the clock gating logic.
4. Do not use clock gating for the master slave flip-flops. Generally, it is normal practice that clock gating logic is used at the slave flip-flop if the clock gating conditions are met. Such design may not perform the desired operation. Use the command **set_clock_gating_exclude** to exclude the master slave flip-flops.
5. While using the clock gating, it is a common practice to use the minimum bus width. The minimum bus width can be of 5 or more. Use the command **set_clock_gating_style_minimum_bitwidth**.
6. In most of the design practices at the RTL level if the procedural 'always' blocks are used and if it consists of 'case' with the 'default' clause or conditional expressions like 'if-else,' then modify the RTL by including the default condition in every 'if-else' statement. Example 1 describes the modification of the procedural block using 'default' as 'else' clause.

Example 1 RTL tweak for the power saving

```
case(a_in)
 2'b00: if (b1_in) c_in =d1_in;
2'b01: if (b2_in) c_in =d2_in;
default : c_in = e1_in;
endcase
The above Verilog RTL can be modified as
case(a_in)
 2'b00: begin
                    if (b1_in) c_in =d1_in;
                     else c_in=e1_in;
           end
2'b01:begin
                   if (b2_in) c_in =d2_in;
                   else c_in =e1_in;
endcase
```

7. If same enable is shared by the multiple register banks, then the power compiler feature can be used to share the clock and enable signal to multiple register bank. This is used to save the overall area. Consider Example 2 shown below, and it has two different procedural blocks then the same clock gating logic can be used for both of the procedural blocks.

Example 2 Use of common clock enable

```
always @ ( posedge clk or negedge reset_n)
begin : block_1
      if (~reset_n)
      data_out <= 1'b0;
     else if (enable)
      data_out<=data_in;
end
always @ ( posedge clk or negedge reset_n)
begin : block_2
      if (~reset_n)
      data_out _1<= 1'b0;
     else if (enable)
      data_out_1<=data_in_1;
end
```

8. Use the simple clocking strategies for the automatic clock gating insertion. If the number of clock domains is minimum, then it gives simplified timing analysis and clock tree synthesis. The lower down modules can have enable signals instead of dividing the clock. Use the **set-don't_touch_network** command to avoid the compilation changes on the clock network. During the multiple step compilation process, this avoids the changes on the clock gating logic.
9. Use the simple set and reset strategies. Complex set and reset strategies may result in the design logic which is prone to issues at the gate-level functional debugging. The care needs to be taken by the designer to have the proper logic partition for synthesis while using the internal set and reset signals.
10. Clock balancing and the clock buffer signal insertion need to be used efficiently to have efficient clock tree synthesis (CTS). CTS tools work by adding or moving the buffers, resizing of cells along the clock tree network to manage the required skew and the insertion delay.

8.4 Switching and Leakage Power Reduction Techniques

There are several techniques used to reduce the power, and few of the commonly used power management techniques are listed in Table 8.2.

Another few important techniques used in the power management at various abstraction levels are listed in Table 8.3.

Table 8.2 Power management techniques

Power management technique	Description
Clock gating and clock tree optimizations	In this technique, the portions of the clock tree which are not used at the instance of time are disabled
Logic restructuring	Use the cone structure to minimize the power. Move the low-switching operations back in the logic cone and high-switching operations up in the logic cone. This technique is used to reduce the dynamic power at gate-level optimizations
Operand isolations	This technique is effective in reducing the power dissipation in the data path of any blocks by using the enable signals
Logic and transistor resizing	Use the downsizing to reduce the leakage current and use upsizing to reduce the dynamic current by improving the slew times
Pin swapping	Use the swapping gate pins to reduce the power. If the capacitance is lower. then the switching can be fast at the gate or pin

Table 8.3 Efficient power management techniques

Power management technique	Description
Multi-V_{th}	Use the multithreshold libraries for the power reduction. Use the high-switching threshold for lesser leakage power, but it reduces the design performance. Use the low-switching thresholds for the higher performance, but it has higher leakage
Multiple supply voltage (MSV islands)	Use the multiple supply voltages for the different design blocks
Dynamic voltage scaling (DSV)	In this technique, the selected blocks can run at different supply voltages according to the design requirements
Dynamic voltage and frequency scaling (DVFS)	This is used to reduce the dynamic power. In this method, the selected blocks of design use the different supply voltages and frequencies on fly
Adaptive voltage and frequency scaling (AVFS)	This can be accomplished by using analog circuits, and in this technique based on the control loop feedback the wide range of voltages is set dynamically
Power gating or power shutoff (PSO)	If the functional blocks are not used, then the selected functional blocks are powered off
Splitting memories	If the memories are controlled by software or the data, then the portions of memories can be spitted into more number of portions. This is effectively used to save the power by shutting off the portion of memories which are not used

8.4.1 *Clock Gating and Clock Tree Optimizations*

This technique is especially efficient in reducing the dynamic power. In most of the application, the power is wasted due to unnecessary toggling of the clock signal. Even the clock trees are the major sources for the larger dynamic power, as they have the larger capacitive load and the switching requires the maximum rate. So if the data is not loaded in the register frequently, then significant amount of power is wasted and this can be saved by using clock gating technique. The clock gating is at the register level or leaf level, and if it is done at the block level, then the entire functional block can be disabled by disabling the clock tree. This reduces the switching and hence reduces the dynamic power.

8.4.2 *Operand Isolations*

This technique is effective in reducing the dynamic power dissipation in the data path of any blocks by using the suitable enable signals. Most of the times, the data path elements are sampled periodically and hence this sampling can be controlled by using the enable inputs. During inactive state of enable signal, the data path inputs can be forced to the constant value to reduce the dynamic power due to lesser switching.

8.4.3 Multiple V_{th}

This technique is effective while optimizing for area, power and speed by using the different threshold voltage. Most of the libraries have the different switching threshold voltages. The efficient EDA tool used for synthesis can be able to use the different library cells of different switching threshold voltages for meeting the area and speed constraints with the lowest power dissipation. Always there is trade-off between the power and speed.

8.4.4 Multiple Supply Voltages (MSV)

In this technique, the different functional blocks operate at the different voltage levels. As the voltage level reduces , the active power is reduced as it is function of the square of the supply voltage. But it can degrade the design performance. While using this technique, it is required to use the level shifters while communicating between the different power domains. If level shifters are not used, then the sampling of the valid signals is an issue.

8.4.5 Dynamic Voltage and Frequency Scaling (DVSF)

Dynamic voltage and frequency scaling is efficient technique to reduce the active power consumption. As discussed in the earlier section, the power dissipation is proportional to the voltage square so lowering the voltage has effect on the power consumption. In this type of technique, depending on the performance and power requirements, the frequency and voltage can be scaled down on the fly and hence it can reduce the power dissipation. This technique is especially effective to optimize the static and dynamic power due to optimization of the frequency and voltage levels.

8.4.6 Power Gating (Power Shut Off)

Power gating or power shutoff (PSO) is one of the effective techniques, and in this technique the design modules which are not used can be switched off using switches. This is one of the powerful techniques used to reduce the leakage power. In most of the industrial applications, the leakage power can be reduced by more than 90% by using the power gating switches. To design this technique, it needs the clear understanding of the power down sequence and use of the isolation cells. It is essential to use the isolation logic with the state retention elements and even level shifters while using the power gating.

8.4.7 Isolation Logic

This is used at the output of power down block to prevent unpowered signals, floating signals from power down block. In the simulation, these signals can be denoted by 'X'. Isolation cells are used between the two power domains and connected between the power off domain and the power on domain. The reason for isolation cell in the two power domain is to isolate the output of blocks before the power switch off state and needs to remain isolated until the power is switched on. In few design scenarios, isolation cells can be used to block levels to prevent the connection to power down logic.

8.4.8 State Retention

During the power off mode, most of the time it is essential to retain the state of registers. The state of the registers is useful during the power recovery. In most of the low-power designs, the state retention power gating flip-flops are used and these flip-flops are called as SRPG. Most of the EDA tool cell libraries are having the SRPG cells.

8.5 Low-Power Design Architecture and Use of UPF

Consider the low-power design architecture which is shown in Fig. 8.7 which we have already discussed at the beginning of the chapter. What we need to have is the understanding of the UPF and the different low-power design strategies which can be useful during the architecture and micro-architecture design.

Unified Power Format (UPF) is the standard used to design electronic systems by considering the power as the feature. The standard is used for low-power ASIC designs. The reasons for using UPF are

1. There is no any method which can support accurate management and distribution of low power at the HDL-level abstraction.
2. Vendor-specific power formats are inconsistent and are prone to bugs due to inconsistent specifications.

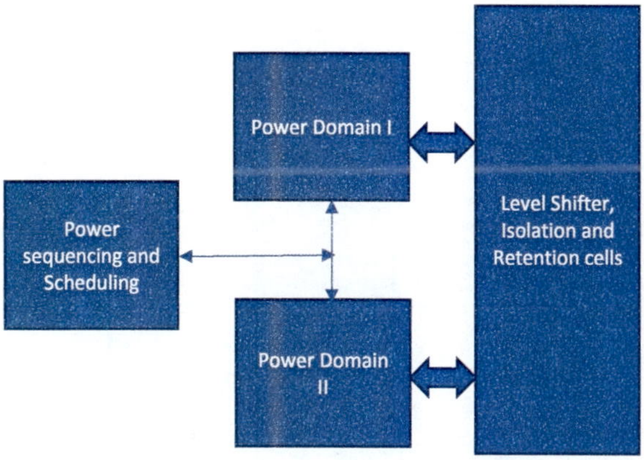

Fig. 8.7 Low-power architecture

3. UPF provides the following and can be used consistently in low-power ASIC designs
   (a) Power distribution architecture
      i. Define the power domains
      ii. Define power switches
      iii. Define power rails
   (b) Power strategy
      i. Creation of power state tables
   (c) Set and map
      i. Isolation
      ii. Retention
      iii. Level shifter
      iv. Switches.

UPF is IEEE 1801 standard and can be used throughout the design flow for power aware design intent. Figure 8.8 describes use of UPF at various stages.

1. **Isolation cells**
   As discussed already, the isolation cells are used at the output of powered down block. The isolation cell can be set by using the UPF command.
2. **Retention cells**
   As discussed already in the above section, the retention cells are used to retain the state of key registers during power off state.
3. **Level shifters**
   Level shifters are used to translate from one voltage level to another voltage level. The translation can be from low to high voltage level or high to low voltage level. Set and map level shifter can be achieved by using the UPF commands.
   The key points to consider for the same are

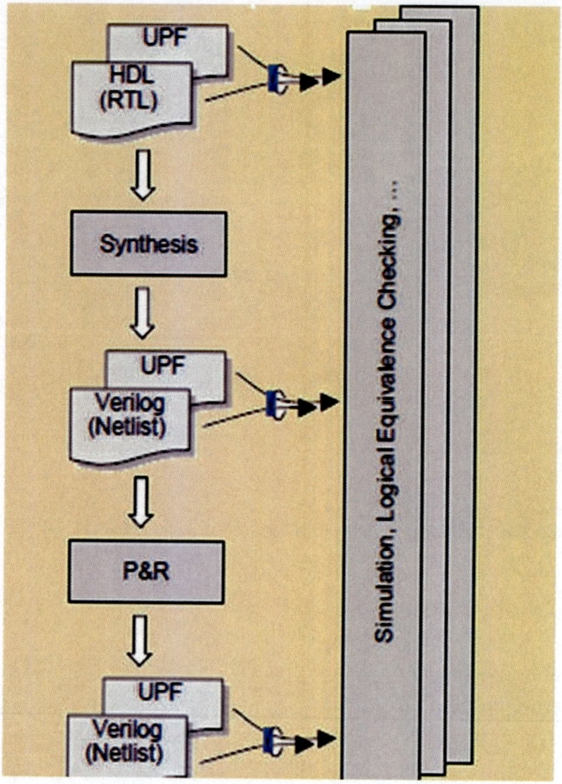

Fig. 8.8 Use of UPF during various design stages

(a) Pick the correct power domain
(b) Select input or output ports or both
(c) Use up shift or down shift rule
(d) Define the location.

4. **Power sequencing and scheduling**
   Specific sequence is generally followed for the power down. The sequence includes isolation, state retention, and the power shutoff. For the power-up cycle, the opposite sequence needs to be followed. During power-up cycle, it is recommended to have the specific reset sequence. Following timing sequence gives information about the power-up/down sequence (Fig. 8.9).

For the multiple clock domains with the different power sequence and the multiple clock gating with few common power control signal, it requires the higher verification efforts to ensure the correct sequencing for the power on and power off.

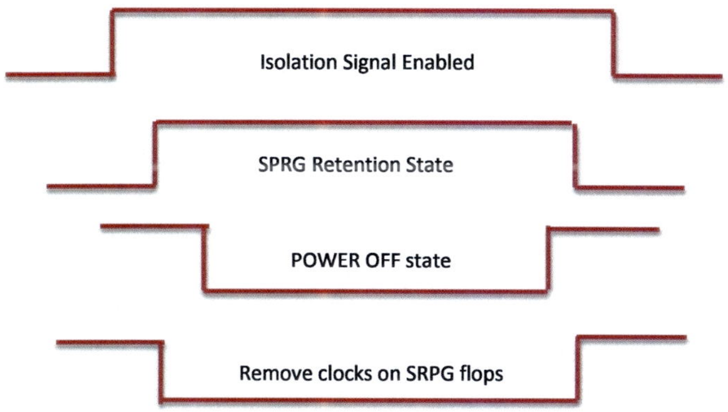

Fig. 8.9 Power sequence

8.6 Chapter Summary

Following are the key highlights to summarize this chapter

1. The power dissipation can be reduced by optimizing for the leakage and dynamic power.
2. Dynamic power can be reduced by reducing the switching activity.
3. The clock gating strategies are useful to reduce the dynamic power.
4. Operand isolation is effective in reducing the dynamic power dissipation in the data path.
5. Dynamic voltage and frequency scaling efficient technique to reduce the active power consumption.
6. Level shifters are used to translate from one voltage level to another voltage level. The translation can be from low to high voltage level or high to low voltage level.
7. The retention cells are used to retain the state of key registers during power off state.
8. Unified Power Format (UPF) is the standard used to design electronic systems by considering the power as the feature. The standard is used for low-power ASIC designs.
9. Power gating or power shutoff (PSO) is one of the effective techniques, and in this technique the design modules which are not used are switched off using switches. This is one of the powerful techniques used to reduce the leakage power.

Chapter 9
Architecture and Micro-architecture Design

The architecture for the ASIC is complex, and it requires the significant amount of experience to finalize and to document the architecture and micro-architecture. The architecture and micro-architecture design are discussed in this chapter and useful during the ASIC design phase.

The important strategies can be the following to develop the architecture of the chip

1. Understanding g of the functionality and block-level representation
2. Single or multiple clocks
3. Power requirements
4. Area and speed requirements
5. Parallelism
6. Pipelining
7. External interfaces
8. Technology node.

9.1 Architecture Design

For any kind of product development based on the ASIC, what we need to understand is the functional specification first and then we need to play around the

1. External interfaces
2. Electrical characteristics
3. Speed, power, and area requirements
4. Mechanical assembly or packaging
5. Design and verification strategies
6. Testing strategies.

The architecture and micro-architecture document plays important role during the design cycle.

V. Taraate, *ASIC Design and Synthesis*,
https://doi.org/10.1007/978-981-33-4642-0_9

With reference to above as per as the functionality of the design is concerned and the requirements of the area, speed, and power, the chapter discusses the architecture and micro-architecture design concepts which can be useful for the complex designs. Consider the video encoder H.264 encoder used to process the HD size image of 1920 × 1080P. Initially , what we will work on the functional blocks of the design, and then we will use the experience to finalize the architecture and micro-architecture for the design.

The important functional blocks are shown in Fig. 9.1.

1. Frame buffers
2. Prediction (Intra or Inter)
3. Storage buffers
4. Quantization and transform (Q&T)
5. Inverse quantization and transform (Q&T)
6. Deblocking filter (DB filter).
   The architect team uses the following to design and finalize the architecture of ASIC!

   (a) Design functionality and understanding of the design application
   (b) Where the design will be used and the constraints associated with it
   (c) Optimization constraints such as speed, power, and area
   (d) Foundry laid rules that is DRC

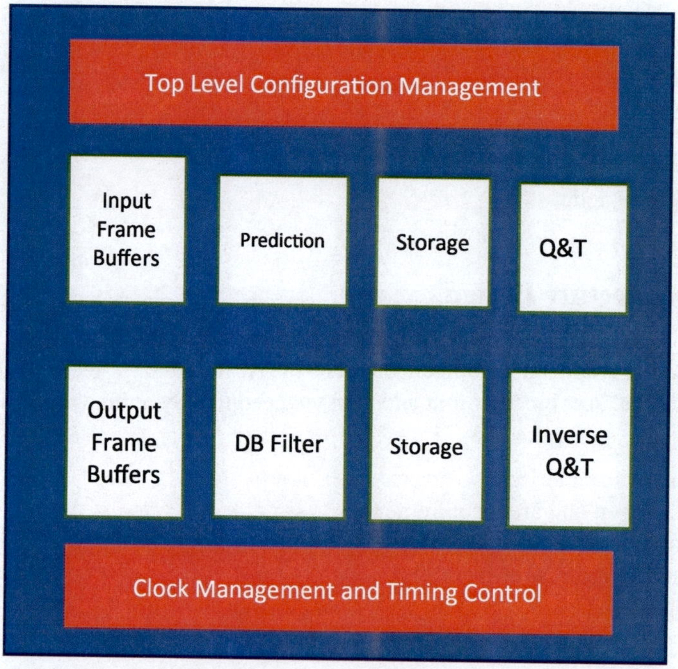

Fig. 9.1 H.264 block-level representation

(e) Requirement of low-power architecture and low-power sequencing
(f) The multiple clock domain designs and strategies to have the different clock groups
(g) The IP requirement during various phases may be hard or soft IPs
(h) Memories and macros requirements
(i) The overall data rate, timing, and clocking requirements for the design
(j) Overall strategies for the hardware and software partitioning of the ASIC design
(k) The testing setup and EDA tools required
(l) The electrical characteristics and interface timing requirements.

All above points are useful to design the architecture and micro-architecture for the chip. By using the functional understanding, the multiple architectures are designed for the ASIC and the best suitable architecture for the design is finalized. The strategies for finalization of the architecture are always dependent on the

1. How best the parallelism and concurrency can be incorporated in the design to have better performance.
2. How the architecture can yield into the better configuration management.
3. How the architecture can allow the design teams to have better initial floor plan to avoid the congestion in the design.
4. How the architecture can give better visibility of the constraints requirement.

9.2 Micro-architecture Design

After the architecture for an ASIC is finalized, the blocks from the architecture are represented in the small blocks or sub-blocks, and it is called as micro-architecture of the design the micro-architecture should be finalized by considering the following!

1. The sub-block functionality and interface strategies
2. The hierarchy of the design
3. The flatten verses hierarchical design and initial gate count estimations
4. Various techniques which can be useful to finalize the block level to constraints
5. Low-power strategies at the block and top level
6. Multiple clock domain interface handling.

9.3 Use of Document During Various Design Phases

The architecture and micro-architecture document used during various phases of the design and these phases are

1. RTL design
2. Synthesis
3. Physical design.

During the RTL design phase, the document is useful to describe the functionality of various blocks using Verilog design. The modular design approach is used due\ring the design, and finally top-level integration is carried out.

During the RTL verification, the verification planning and the verification architecture are used, and it is out of scope as per as discussion is concerned.

During the block and top-level synthesis also with the Verilog source files if the architectures know to the synthesis team them\n during the performance improvement, it can be used as a better tool.

Even the better architecture and micro-architecture can be used to have the better strategies for the initial floor plan and placement of the functional blocks.

9.4 Design Partitioning

The architecture and micro-architecture document should also give the information about the design partitioning as per as functionality of the design is concern. The following should be documented to have the better outcome of the ASIC design!

1. **Hardware and Software Partitioning**: The functional blocks need to be designed using the Verilog and the blocks need to be designed using the software should be documented and are one of the factors to establish the better communication and synchronization for the ASIC or SOC blocks.
2. **Partitioning at the Logic Level**: To have the better performance, the partitioning strategies should be documented at the logic level. For example, complex processors functional blocks and interfaces can be partitioned at the top level to have the modular design approach.
3. **Multiple Clock Domain Designs**: Partitioning of the design by combining the functionality depending on the clock groups can result into the better synthesis and in turn better floor plan.
4. **Partitioning for Low-Power Aware Architectures**: Partitioning of the design by considering the power requirements can result into the better power optimizations during the design.
5. **Partitioning Analog and Digital Domains**: For the ASIC designs which uses the analog and digital blocks, the better strategy is to partition the design into the analog and digital domain. Have the full-custom design flow for the analog blocks and the semi-custom ASIC flow for the digital design.

9.5 Multiple Clock Domains and Clock Grouping

Let us discuss the design which needs to have the multiple clocks. The processor and associated logic works on the clk1. Memory controller works on clk2, and the DSP processor with the H.264 encoder works on clk3. The design has three clocks, and Table 9.1 gives information about the clock frequency (Fig. 9.2).

Table 9.1 Clock frequencies at various clock domains

Clock domain control	Frequency in MHz	Description
clk1	500	The clock domain one operating at frequency of 500 MHz
clk2	666.66	The clock domain two operating at frequency of 666.66 MHz
clk3	250	The clock domain three operating at frequency of 250 MHz

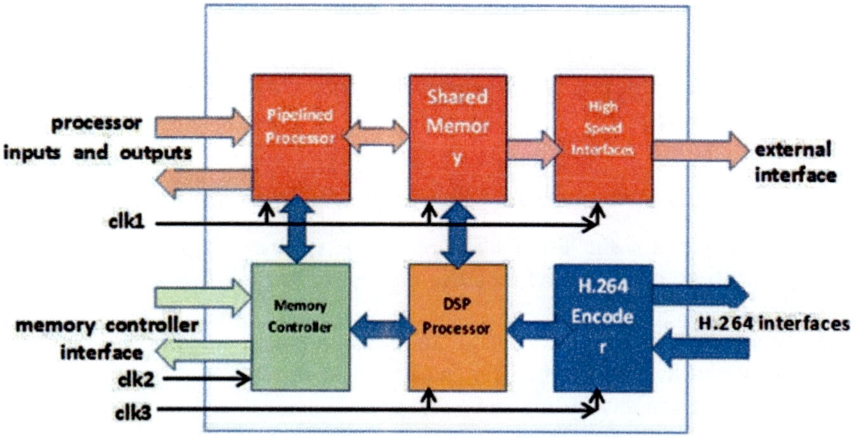

Fig. 9.2 Multiple clock domain design architecture

For such kind of designs, the partitioning by considering various clock domains can play especially important role. It will be useful during the RTL design phase while deploying the synchronizers in the control and data path.

9.6 Architecture Tweaking and Performance Improvement

Architecture and micro-architecture tweaks can incur significant amount of time and budget in the ASIC design cycle. Even the manufacturing processes will be delayed if major architecture and micro-architecture changes are required.

Let us try to understand the scenarios during the design which may require major tweaks in the architecture and in turn all the design phases.

1. **Specification Additions**: If client suggests incorporating the additional functionality in the design, then it becomes very time consuming to tweak the present architecture. It adds the significant amount of the time and budget and elongates the overall chip development cycle. Reason may be the additional logic required

for debugging and for the test of the ASIC or need to have the low-power aware design.

2. **The Architecture is Having Parallelism**: If the architecture is having the more parallel blocks which executes concurrently and during synthesis if the team faces issues in the area optimization, then the architecture and micro-architecture tweaks can be carried out. But prior to this, it is always recommended to work on the RTL tweaks and tool-based i\optimization techniques.
3. **Performance is Not Met During the Synthesis**: Initially, it is recommended to carry out the RTL tweaks, and if design doesn't meet the performance, then carry out the tweaks in the architecture.

9.7 Strategies for the Micro-architecture Design of Processor

Let us consider the 32-bit processor which has the following specifications

1. It should perform the arithmetic operations such as addition, subtraction, multiplication, division, and modulus on signed, unsigned, and floating-point numbers.
2. It should perform the logical operations on 32-bit binary numbers.
3. It should perform the data transfer and branching operations.
4. It should perform the shifting and rotate operations.
5. The external interfaces can be
   (a) IO interfaces
   (b) Serial IO
   (c) High-speed interfaces
6. It should have the internal memory storage of 64 KB.
7. The processor should have the interrupt controller.
8. The processor should have two clock domains and should use clk1 and clk2, respectively.

Consider that these specifications are extracted from the requirements, and let us try to use these to have better architecture and micro-architecture.

1. **Multiple Clock Domain Groups**:
   Clock domain 1: It is controlled by the clk1 and the functional blocks of this clock domain are
   (a) ALU
   (b) Internal memory
   (c) Interrupt controller
   (d) Pointers and counters
   (e) Serial IO
   (f) IO interfaces.

In the architecture clock, domain 1 blocks are indicated by the yellow color.

Clock domain 2: It is controlled by the clk2, and the functional blocks of this clock domain are

(a) Floating-point unit
(b) High-speed interfaces.

In the architecture clock, domain 2 blocks are indicated by the white color.

2. **Processor Engine**
   As stated in the specification extraction document, the processor core performs various operations on the signed and unsigned number and floating-point numbers so the better strategy is to have the dedicated block for the general-purpose operations and floating-point operations (Fig. 9.3).
   ALU: performs the operations on the signed and unsigned numbers.
   Floating-Point Engine: used to perform the operations on the floating-point numbers.
   Then let us try to have the dedicated memory block of 1 MB may be have partitioning according to the address ranges so that various functional units can perform the read and write operations.
3. **Memories**: To store the internal data, the processor needs to have the internal memory and can be shared between the general-purpose ALU and the floating-point engine. If we have the multiple clock domain design, then better way is to have the separate memory for the general-purpose processor and floating-point engine.
   With reference to the specification, the 64 KB memory is divided into two blocks of 16 KB and 48 KB, respectively (Fig. 9.4).

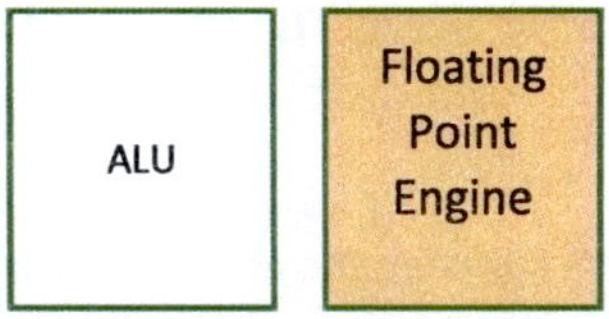

Fig. 9.3 Processing engine

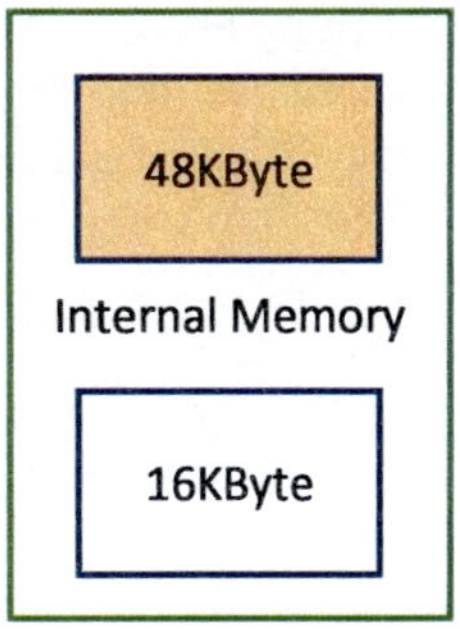

Fig. 9.4 Memories

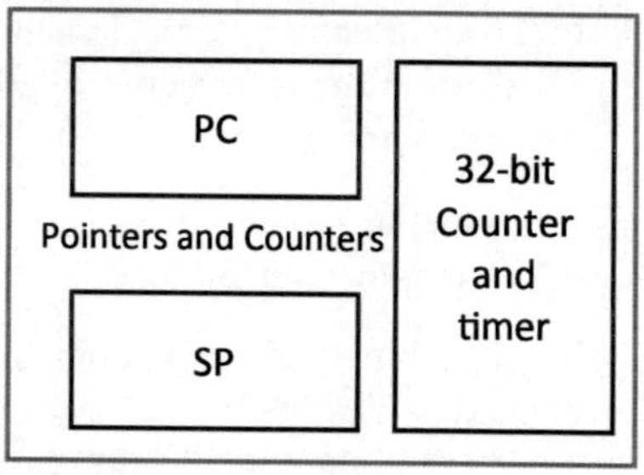

Fig. 9.5 Counters and pointers

4. **High-Speed Interfaces**: To exchange the data from the external memory and IO after performing the floating-point operations, the architecture need to have the high-speed interfaces. These high-speed interfaces are designed to have low latency and minimum interconnect delays.
5. **Pointers and Counters**: During the general-purpose processing of the data, the result may need to be stored in the reserved area of memory so the design may need the stack pointer and to fetch the instruction and the data from the external memory the design needs the program counter. The stack and program counter are 16-bit for the architecture (Fig. 9.5).
   The 32-bit counter and timer are used as dedicated timer and counter during the counting applications.
6. **IO and Communication Blocks**: To communicate with the external devices such as serial and parallel, the processor architecture should have the dedicated blocks. These blocks are

   - **IO Interfaces**: For the 32-bit data transfer dedicated high-speed IOs to exchange the 32-bit of the data between the IO devices and processor
   - **Serial IO**: The serial devices can communicate with the general-purpose processors using the serial IOs.

7. **Interrupt Controller**: The architecture provides the dedicated block to process the edge and level-sensitive interrupts. The interrupt controller can halt the current execution for the valid interrupt.
8. **Clock Management and Timing Control**: The clock management and timing control block to distribute the clock with uniform clock skew.
9. **Configuration Management**: The configuration and test management to carry out the initial test and to manage interaction between the processor and the system (Fig. 9.6).

The document should consist of the information regarding the following

1. Initial floor plan and initial area estimation
2. Information about the timing and power
3. Information about the interfaces between various functional blocks
4. Information about the design partitioning
5. EDA tool requirements

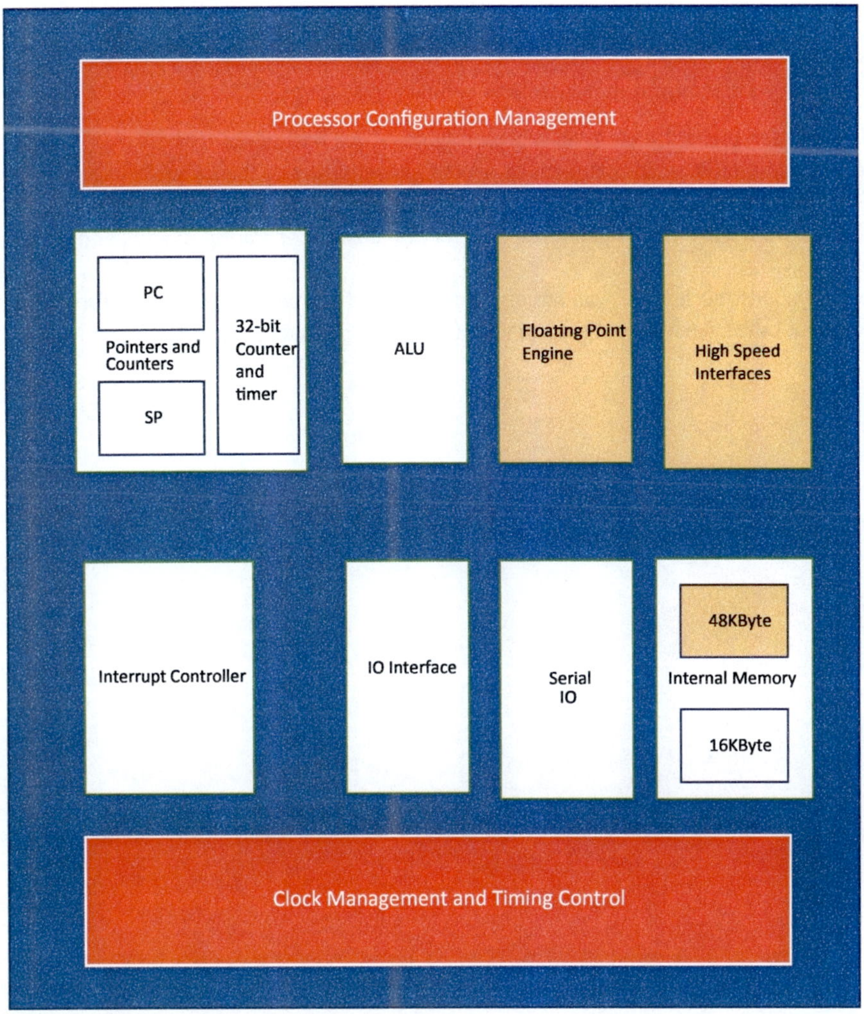

Fig. 9.6 Sub-block-level representation for processor

6. IP and memory requirements
7. Overall clocking strategies
8. Reset policies
9. Overall block, top, and chip-level timing.

For more information about the architecture and micro-architecture, refer case studies documented in Chaps. 17, 19 and 20.

9.8 Chapter Summary

Following are the important points to conclude the chapter.

1. Architecture is block-level representation of the design.
2. Micro-architecture is sub-block-level representation of the design.
3. It is recommended to have the better strategies for the design partitioning.
4. Partition the design during the architecture design for multiple clock and power domains.
5. The better architecture and micro-architecture document should give information about the interface and timing with the interdependability of the blocks.

Chapter 10
Design Constraints and SDC Commands

As discussed during Chaps. 5–9, the important ASIC design constraints are classified into two categories, and mainly they are

1. Optimization constraints
2. Design rule constraints (DRC).

The optimization constraints are speed and area as per as the ASIC logic design is concern. During the physical design, we need to optimize the design for the area, speed, and power. The better power planning depending on the desired technology node and the strategies is always helpful to get the layout of the chip.

The DRC are the foundry laid rules and mainly the transition, fanout and the capacitance.

The constraints can be used to optimize the design during various synthesis phases during the logical and physical synthesis.

These constraints are at the block, top, and chip level of the design. Consider the architecture of the processor as shown in the Figure and the block-level constraints can be specified for various functional blocks which are ALU, floating-point engine, high-speed interfaces, etc. The top-level constraints will be used during the synthesis, and they are for the integration of all the functional blocks.

If the block-level constraints are met, it doesn't mean that the design will meet the top-level constraints. The chip-level constraints for the clean layout need to be met during the physical design (Fig. 10.1).

For the synthesis of the processor, following can be the better strategies

1. Perform the synthesis for the different clock groups.
2. Use the bottom-up synthesis and extract the block-level constraints.
3. Optimize the design during block-level synthesis to meet the area and speed.
4. Specify the top-level constraints.

Design optimization constraints are speed, area, and power.

V. Taraate, *ASIC Design and Synthesis*,
https://doi.org/10.1007/978-981-33-4642-0_10

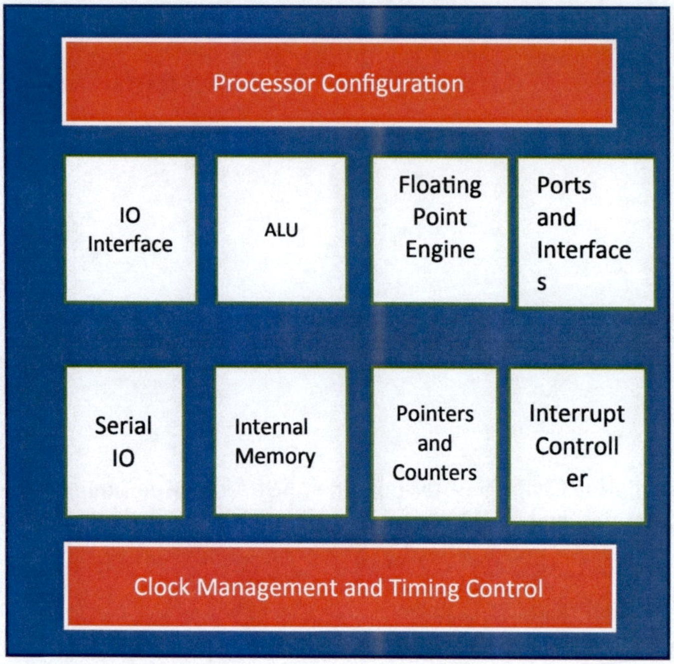

Fig. 10.1 Processor top-level architecture

5. Perform the top-level synthesis and optimize the design to meet the top-level constraints.
6. Use the strategies to tweak the RTL, architecture if the constraints are not met.

10.1 Important Design Concepts

10.1.1 Clock Tree

The clock tree synthesis is carried out during the physical design flow, and during the logic design flow we do not have the information about the clock distribution. That is, we will try to use the Synopsys DC setup with the statistical data available to specify the clock and clock latency.

10.1.2 Reset Tree

The design which has the multiple resets for the initialization of the functional blocks need to be synchronized with the main reset that is master reset. The reset trees can be useful to avoid the metastable output if the reset is generated during the active edge of the clock.

The important parameters to consider are the

1. Reset recovery time.
2. Reset removal time.

10.1.3 Clock and Reset Strategies

The following strategies can be helpful during the logic design as per as reset and clocks are concerned.

1. For multiple clock domains, use the synchronizers in the data and control path which is already discussed in Chap. 7.
2. Use the statistical data to introduce the clock latency and specify the setup and hold uncertainty during logic synthesis.
3. Hand instantiate the clocks during the logic design.
4. Use the reset synchronizers to synchronize the reset with the master reset.

10.1.4 What Impacts on Design Performance?

The ASIC design should meet the optimization constraints for speed and area. The power constraints and DRC we will use are during the physical design. The following are important points need to address during the synthesis

1. **Block-level constraints**: For the complex ASIC designs if we consider the multiple functional blocks or Ips, then the block-level constraints should be specified. The block-level constraints for the functional block should meet. For example, the processor logic operates at the 250 MHz operating frequency but the overall chip works at the 500 MHz. In such scenario, the overall uncertainty for the setup and hold is different as compared to top-level constraints. So, the block-level Tcl script should be used during the block-level synthesis.
2. **Top-level constraints**: For the bottom-up synthesis after performing the synthesis of all the functional block, the top-level integration is carried out. The top-level constraints need to be specified for the specific clock groups and mainly in the Tcl script, the following commands should be used
   (a) Clock latency information
   (b) Input delay
   (c) Output delay
   (d) Setup uncertainty
   (e) Hold uncertainty.

If the block-level constraints are met, then it is not guaranteed that the top-level constraints will meet. Few of the reasons may be

1. The additional delays incurred if the design partitioning is not at sequential boundaries.
2. The data arrival is fast, and the design has hold violations.
3. The data arrival is slow, and the design has setup violations.
4. If during the synthesis, there are timing exceptions due to multicycle and false paths.
5. The data integrity due to poor synchronization strategy.
6. If the design has hierarchy and the DC not able to optimize the glue logic. In such scenarios, the design needs to be flattened to improve the optimization.

10.2 How to Interpret the Constraints

The important constraints which need to be specified for the block and top-level synthesis with Verilog files are area, speed, and power. Let us rule out the power as power optimization is not carried out using DC. As a designer and synthesis team member, our goal is to have the functional understanding of the design and overall area and speed requirements for the design.

10.2.1 Area Constraints

During the logic synthesis, the area is due to the logic and macros used. The standard cell information is available in the library, and the specific macros are required to have the lower-level abstraction of the design. The overall area optimization can be carried out during the

1. **RTL design**: Using the few concepts like resource sharing, resource allocation, eliminating dead zones, using parenthesis and groping.
2. **Synthesis**: By using tool specified directives and using the area optimization commands, the area can be optimized.

10.2.2 Speed Constraints

Speed is especially important factor as it decided the overall performance of the design. The speed constraints for the design need to be worked out depending on the statistical data available in the library for the specific technology node, and these constraints should be met. As the actual placement and routing information is not available during the logic synthesis, the goal is to have close look to eliminate the setup violations for the block- and top-level design. The synthesis and STA team need to specify the following

1. Clocks
2. Clock latency
3. Setup and hold uncertainty
4. Input and output max and min delays
5. Specify the multicycle paths
6. Specify the false paths.

10.2.3 Power Constraints

The power is another constraint, and during the power planning we specify the constraints as leakage and dynamic power. To have the low-power aware architectures and designs, we will use the Unified Power Format (UPF) at various design stages.

Following can be few of the strategies to optimize for the power

1. **Architecture design**: Have the low-power architecture design and have the strategies for power sequencing and power shutdowns.
2. **Use the low-power cells**: Use the low-power cells during design but the designer needs to have better understanding of the cell characterization as use of these cells has significant impact on the speed of the design.
3. **RTL design**: During the RTL design, use the clock gating cells to reduce the dynamic power.

10.3 Issues in the Design

Following are the important challenges during the ASIC synthesis

1. Trimming of the logic
2. Unconnected ports and nets
3. The block-level speed constraints met but at the top-level design fails
4. Design has the missing block level connectivity although the RTL verification is successful.

10.4 Important SDC Commands Used During Synthesis

This section discusses the important DC commands used during the synthesis and useful to specify the constraints. Refer Chaps. 11, 12, and 13 to understand the synthesis and optimization with the design scenarios.

10.4.1 Synopsys DC Commands

Few of the SDC command used during the ASIC synthesis are documented in this section.

1. **Reading the design**

    read –format <format_type> <filename>

    The above command is used to read the design.

> For example to read the processor top use the following SDC command
>
> ***read-format verilog processor.v***

2. **Analyze the design**

    analyze –format <format_type> <list of file names>

    Used to analyze the design. It is used to report the syntax errors and to perform the design translation before having the generic logic. The generic logic is part of the synopsys generic technology-independent library. The components are named as GTECH. This logic is unmapped representation of the Boolean functions.

> The command used to analyze the processor,v file is
>
> **analyze –format verilog processor.v**

3. **Elaborate the design**

    elaborate –format <list of module names>

    Used to elaborate the design and can be used to specify the different architectures during elaboration for the same analyzed design.

> The command used to elaborate design is
>
> **elaborate –library work processor**

It is essential to understand about the difference between the ***read*** and ***analyze***, ***elaborate*** command. The following are key highlights:

1. The analyze and elaborate are used to pass required parameters while elaborating the design.
2. The read is used while entering for the pre-compiled designs or netlists in DC.
3. Using analyze and elaborate commands, the different architectures can be specified during elaboration for the same analyzed design.
4. The read command doesn't allow the use of the different architectures.

10.4.2 Checking of the Design

After the design has been read using the DC, the check_design is used and used to check the design problems like shorts, opens, multiple connections, and instantiations and the no connections.

check_design

The command used to check the errors in the design
check_design

10.4.3 Clock Definitions

The clock needs to be specified using the command create_clock, and this is used as reference clock during the timing analysis. The following example describes the clock definition using the create_clock command.

create_clock –name <clock_name> -period <clock_period> <clock_pin_name>

The command is used create the clock for the design which is used as reference clock during timing analysis. If design doesn't have clock, then it will be treated as virtual clock.

The 500MHz clock with 50% duty cycle is created using create_clock and specified as

create_clock –name clock -period 2 processor_clock

Clock Having Variable Duty Cycle

If designer wish is to use the clock with variable duty cycle having rising edge at 0.5 ns and clock period of 2 ns, then the create_clock command can be modified as

create_clock –name clock - period 2 –waveform {0.5,2} –name processor_clock

Virtual Clock

If the design doesn't have the clock pin, then the virtual clock is created using following commands.

This command generates virtual clock of frequency 500 MHz with 50% duty cycle.

create_clock –name clock -period 2

This command generates virtual clock of frequency 500 MHz with variable duty cycle having rising edge at 0.5 ns and falling edge at 2 ns.

create_clock –name clock -period 5 –waveform {0.5,2}

10.4.4 Skew Definition

As discussed in Chaps. 5 and 6, the skew is the difference between the arrivals of the clock signal. If clock at the source flip-flop is delayed with reference to the destination flip-flop, then the skew is called as negative clock skew and useful for the hold. If clock at the destination flip-flop is delayed as compared to the source flip-flop, then the skew is called as positive clock skew and useful for the setup. The reason being as clock at the destination flip-flop is delayed, and the data can arrive late by the margin of the skew.

The design compiler will not be able to synthesize the clock tree, and so to overcome the problem the clock skew is used to specify the delay!

The following command used by design compiler to specify clock skew for the design

set_clock_skew –rise_delay <rising_clock_skew> -fall_delay <falling_clock_skew> <clock_name>

> This command used to specify the clock skew for the design and described as:
>
> **set_clock_skew –rise_delay 2 –fall_delay 1 master_clock**

10.4.5 Defining Input and Output Delay

The input and output delay can be specified by using set_input_delay and set_output_delay commands, respectively. The command used to specify the input and output delay is specified below.

set_input_delay –clock <clock_name> <input_delay> <input_port>

Used to define the input delay.

> To define 1ns delay with reference to clock, the command can be used as
> **set_input_delay –clock master_clock 1 data_in**

set_output_delay –clock <clock_name> <output_delay> <output_port>

Used to define the output delay.

> To define 1ns delay with reference to clock To define 1ns delay with reference to clock, the command can be used as
> **set_output_delay –clock master_clock 1 data_out**

10.4.6 Specifying the Minimum (min) and Maximum (max) Delay

The input and output delays can be specified as min or max depending on the design needs.

Maximum Input Delay

set_input_delay –clock <clock_name> -max <delay> <input_port>

Used to define the max input delay.

To define 2ns delay with reference to clock, the command can be used as
set_input_delay –clock master_clock –max 1 data_in

Minimum Input Delay

set_input_delay –clock <clock_name> -min <delay> <input_port>

Used to define the minimum delay.

To define 1ns delay with reference to clock, the command can be used as
set_input_delay –clock master_clock –min 1 data_in

set_output_delay –clock <clock_name> -max <delay> <output_port>

Used to define the maximum output delay.

To define 2ns delay with reference to clock, the command can be used as
set_output_delay –clock master_clock –max 2 data_out

Minimum Output Delay

set_output_delay –clock <clock_name> -min <delay> <output_port>

Used to define the minimum output delay.

To define 1ns delay with reference to clock, the command can be used as
set_output_delay –clock master_clock –min 1 data_out

10.4.7 Design Synthesis

The compile command is used to perform the design synthesis. As discussed in previous section, we need to have the design constraints, library, and Verilog files as inputs to the synthesis tool. The design synthesis can be performed using the different efforts levels like low, medium, and high.

The compile command is specified as

compile –map_effort <map_effort_level>

compile –map_effort medium

The command for the medium effort level can be.

10.4.8 Save the Design

The write command is used to save the design. The designer can save the synthesis output in the Verilog (.v) or database (.ddc) format. The command can be specified as shown:

write –format <format_type> -output <file_name>

The command used to save the netlist in Verilog format is specified as :
write –format verilog -output processor_netlist.v

10.5 Constraint Validation

The important commands used to validate the design are listed in Table 10.1.

Table 10.1 Constraint validation

Command	Description
check_design	*Used to check for the design consistency and reports the unconnected nets, ports, etc.*
check_timing	*Used to verify the timing*

10.6 Commands for the DRC, Power, and Optimization

Important commands used to specify design rules, power, and optimization constraints are listed in Table 10.2.

Table 10.2 DRC, power, and optimization definition

Command	Type	Description
set_max_transition	*DRC*	*Used to define the largest transition time*
set_max_fanout	*DRC*	*Used to set the largest fanout for the design*
set_max_capacitance	*DRC*	*Used to set the maximum capacitance allowed for the design*
set_min_capacitance	*DRC*	*Used to set the minimum capacitance allowed for the design*
set_operating_conditions	*Optimization constraints*	*Used to set the PVT conditions as it affects on timing*
set_load	*Optimization constraints*	*Used to model load on output port*
set_clock_uncertainty	*Optimization constraints*	*Used to define the estimated network skew*
set_clock_latency	*Optimization constraints*	*Used to define the estimated source and network delays*
set_clock_transition	*Optimization constraints*	*Used to define the estimated input skew*
set_max_dynamic_power	*Power constraints*	*Used to set the maximum dynamic power*
set_max_leakage_power	*Power constraints*	*Used to set the maximum leakage power*
set_max_total_power	*Power constraints*	*Used to set the maximum total power*
set_dont_touch	*Optimization constraints*	*It is used to prevent the optimization of mapped gates*

10.7 Chapter Summary

Following are the important points to conclude this chapter:

1. The design constraints are optimization and design rule constraints.
2. The synthesis is the process to get the lower-level design abstraction from the higher level.
3. Synthesis tool uses the Verilog files, library, and the constraints as inputs.
4. The output from the synthesis tool is gate-level netlist.
5. The constraints for the block-level and top-level design should be documented in the separate Tcl file.
6. The Synopsys DC does not optimize for the power.
7. During the logic synthesis, the goal is to optimize the design for the area and speed.

Chapter 11
Design Synthesis and Optimization Using RTL Tweaks

The synthesis is the process to get the lower level of design abstraction. If we have the design at switch level or device level, then it is lowest level of abstraction of the design. We specify the functionality of the design using the Verilog, and the design needs to be mapped and routed during the physical design flow. For such requirements, the design goes through various design phases using the sophisticated design tools.

The synthesis is performed at various levels

1. **Logic Synthesis**: RTL design translation into the gate-level netlist. This uses the Verilog files, library and the constraints.
2. **Physical Synthesis**: The gate-level netlist generated from the logic synthesis is translated to have the layout that is at the physical level. The constraints used during the physical design are the top-, chip-level optimization constraints. The flow uses the design rules which are technology specific.

11.1 ASIC Synthesis

The process of getting the lower level of abstraction from the RTL design is called as logic synthesis. The output from the synthesis tool is the gate-level netlist. The EDA tool uses the Verilog files, design constraints library as an input to generate the gate level netlist as an output.

The ASIC synthesis tool uses the inputs as

- Vcrilog files
- Library
- Constraints.

The synthesis tool output is

For better design outcome the design synthesis with optimization goals need to be carried out!.

V. Taraate, *ASIC Design and Synthesis*,
https://doi.org/10.1007/978-981-33-4642-0_11

- Gate-level netlists which can be stored in the Verilog (.v) or database (.ddc) format.

The popular synthesis tools in the industry are

- **Synopsys Design Compiler** which is popular as DC.
- **Cadence RTL Compiler** which is popular as RTL compiler.
- **Role of Synthesis Tool**: The synthesis tool uses the Verilog files, constraints, and the libraries to get lower level of abstraction for design during logic synthesis. That is it is used to get the gate-level netlist. Synthesis tool tries to meet the block and top-level constraints by calculating the cost of various implementations.
- **Gate-Level Netlist**: The gate-level netlist is the structural description using the standard cells.
- **Gate-Level Verification**: The gate-level netlist is verified for the functional correctness of the design, and this is called as gate-level verification.

11.2 Synthesis Guidelines

To have the better performance, use the following guidelines during the ASIC and FPGA synthesis.

1. **Use of the Naming Convention**: Use the naming conventions for all the input and output ports. Use the naming convention specified in Table 11.1.
2. **Partitioning**: Partition the design at the sequential boundaries for better timing and performance. For example, use the registered outputs and inputs.
3. **RTL-Level Strategies**: Few of the strategies useful during the RTL design
   (a) Within the always procedural block use the blocking assignments (=) to infer the combinational or glue logic and the non-blocking assignments (<=) to infer the sequential logic.
   (b) Do not mix blocking and non-blocking assignments.
   (c) Avoid latches by using default in the ***case*** construct.
   (d) Use the ***else*** clause to infer the unintentional latches while using the if-else construct.
   (e) To have complete sensitivity list, use the ***always@****.

Table 11.1 Naming convention

Name of signal or port	Naming convention description
Master clock	It can be named as master_clk
Inputs	Use the inputs as a_in, b_in, data_in
Outputs	Use the outputs as y_out, q_out, data_out
Active low asynchronous reset	Use reset_n
Active low synchronous reset	Use reset_sync_n
Bidirectional signals	Use data_io

4. **Avoid Oscillations**: Avoid the combinational looping in the design as they have oscillatory behavior.
5. **FSM-Based Designs**: Try to optimize the FSM for the better timing and performance using the separate always procedural block for the state register, next state, and output logic. Use the data path and control path as separate module for clean timing and glitch-free design. For more details refer Sect. 11.3.
6. **Avoid Hierarchy in the Combinational Design**: For better synthesis optimization, avoid hierarchy in the combinational design.

11.3 FSM Design and Synthesis

Consider the input string as continuous data having combination 10001010010101001 1-----. Now for the overlapping sequence of the 101, the Mealy machine needs three states. The output for the non-overlapping sequence is given by 000000100001010000---. The state machine is shown in Fig. 11.1.

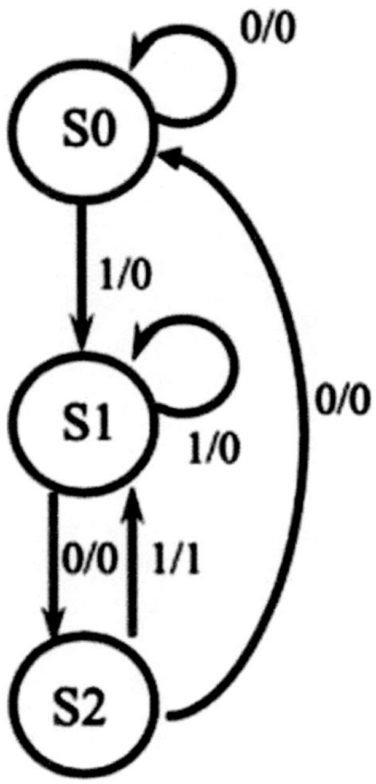

Fig. 11.1 Mealy state machine for overlapping sequence

For the state machine, the description using the Verilog constructs is described in Example 1.

Example 1 Description using Verilog for overlapping Mealy machine

```
/////////////////////////////////////////////////////////////////////////////////////////


module mealy_machine(input  clk, reset_n, data_in, output reg
data_out);
parameter s0=2'b00;
parameter s1=2'b01;
parameter s2=2'b10;
reg [1:0] present_state, next_state;
//state register logic
always @(posedge clk or negedge reset_n)
begin

 if (~reset_n)
present_state<= s0;
 else
present_state<= next_state;

end

// next_state_logic
always@*
begin
case (present_state)

s0 : if ( data_in)
next_state = s1;
   else
next_state = s0;
```

```
s1 : if ( ~data_in)
next_state = s2;
     else
next_state = s1;

s2 : if ( data_in)
next_state = s1;
     else
next_state = s0;
default : next_state=s0;
endcase
end
//output logic
always@*
begin
case (present_state)
s0 : data_out = 0;
s1 : data_out = 0;
s2 : if (data_in)
data_out = 1;
    else
data_out=0;
default : data_out=0;
endcase
end
endmodule
////////////////////////////////////////////////////////////////////////////////
```

The synthesis result is shown in Fig. 11.2 and has the next state logic, state registers and output combinational logic. As shown output is function of the present state and input. As compared to Moore state machine, the logic inferred uses more elements in the output combinational logic.

11.4 Strategies for the Complex FSM Controllers

The following are few of the important strategies useful during the design of the FSM controllers

1. **FSM Description**: Use the multiple procedural blocks to describe the FSM. Use the state register logic, next state logic and output logic.
2. **Glitch-Free Output**: To avoid the glitches in the FSM designs, use the register output and try to have the combinational logic optimization for the better performance.
3. **Encoding**: Use the one-hot encoding for the clean and better timing performance if the area is not an issue.
4. **Data and Control Paths**: Try to have the separate data and control path for the FSM controllers. During the data path and control path synthesis try to optimize for the late arrival signal logic using the architecture and RTL tweaks.
5. **Tool-Based Optimization**: Use the FSM compilers to extract the states and optimize for the controller design for the better area and timing.

11.5 How RTL Tweaks Are Useful During Synthesis?

As ASIC designs are complex, we will try to use the modular approach by partitioning the design at sequential boundaries. The better partitioning is helpful during the RTL design phase and even during the initial floor plan stage. As discussed in the Chap. 3, we can have the strategies during the RTL design so that we will be able to get the better design performance.

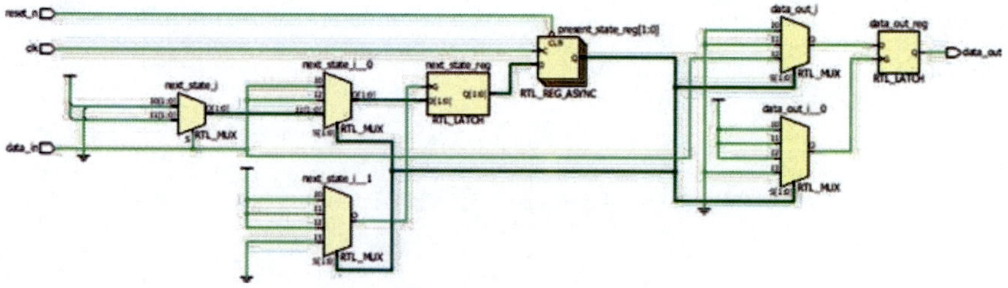

Fig. 11.2 Synthesis result for the overlapping Mealy machine

Table 11.2 Arithmetic operation Table 11.1

Control input (*sel_in_0*)	Operation	Description
1	$y1\_out = a\_in + b\_in$ $y3\_out = a\_in * b\_in$	Perform the addition, multiplication on a_in, b_in
0	$y1\_out = c\_in + d\_in$ $y3\_out = c\_in * d\_in$	Perform the addition, multiplication on c_in, d_in

Table 11.3 Arithmetic operation Table 11.2

Control input (*sel_in_1*)	Operation	Description
1	$y2\_out = e\_in + f\_in$ $y4\_out = e\_in * f\_in$	Perform the addition, multiplication on e_in, f_in
0	$y2\_out = a\_in + b\_in$ $y4\_out = a\_in * b\_in$	Perform the addition, multiplication on a_in, b_in

To have better understanding, consider the design which has the following functionality (Tables 11.2 and 11.3).

The RTL description using the if-else construct is shown in Example 2, and during the design the team has not considered for the performance of the design.

Example 2 RTL without any optimization strategies

```
/////////////////////////////////////////////////////////////////////////

module area_without_optimization (
                    input a_in,b_in, c_in, d_in, e_in, f_in,
                  input sel_in_0, sel_in_1,
                  output reg y1_out, y2_out,
                  output reg [1:0] y3_out, y4_out);

always @ *
begin

 if ( sel_in_0)
 begin
        y1_out = a_in + b_in;
         y3_out = a_in * b_in;
  end
  else
  begin
```

```
            y1_out = c_in + d_in;
           y3_out = c_in * d_in;
   end
 end

 always @ *
 begin
   if ( sel_in_1)
   begin
           y2_out = e_in + f_in;
           y4_out = e_in * f_in;
    end
    else
    begin
            y2_out = a_in + b_in;
           y4_out = a_in * b_in;
    end
end
endmodule
/////////////////////////////////////////////////////////////////////////////////
```

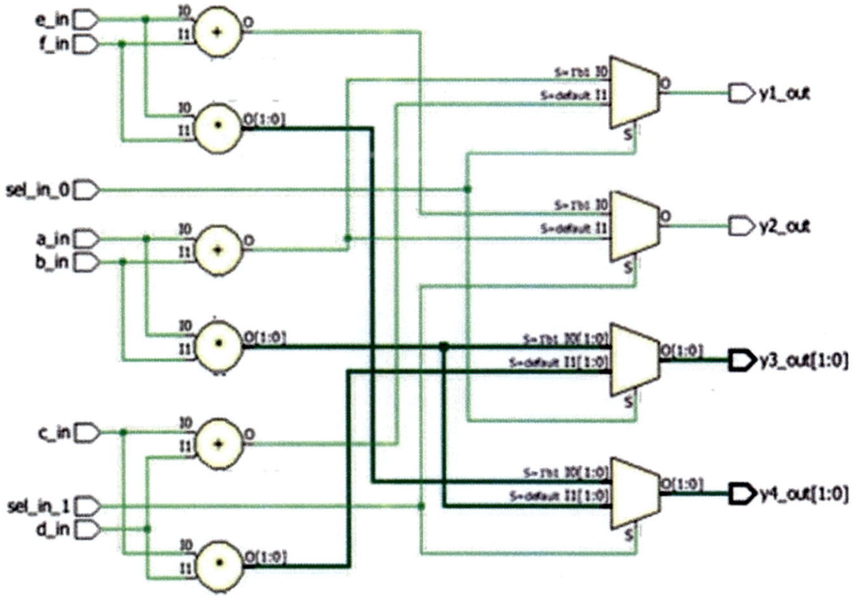

Fig. 11.3 RTL schematic for the design without area optimization

During the ASIC synthesis, the design infers the combinational logic and following are the performance issues for the inferred logic.

1. If the select inputs are late arriving, then the data is unnecessary available at the arithmetic resource. All the arithmetic resources perform the operations at a time and the design does not demand for the same.
2. The area and speed can be improved for such type of design using the sharing of common resources which are adder and multiplier (Fig. 11.3).

Table 11.4 Strategies for resource sharing

Control input (*sel_in_0*)	tmp1	tmp2	Operation	Description
1	*a_in*	*b_in*	*y1_out* = *tmp1* + *tmp2* *y3_out* = *tmp1* * *tmp2*	Perform the addition, multiplication on a_in, b_in
0	*c_in*	*d_in*	*y1_out* = *tmp1* + *tmp2* *y3_out* = *tmp1* * *tmp2*	Perform the addition, multiplication on c_in, d_in
Control input (*sel_in_1*)	tmp3	tmp4	Operation	Description
1	*e_in*	*f_in*	*y2_out* = *tmp3* + *tmp4* *y4_out* = *tmp3* * *tmp4*	Perform the addition, multiplication on e_in, f_in
0	*a_in*	*b_in*	*y2_out* = *tmp3* + *tmp4* *y4_out* = *tmp3* * *tmp4*	Perform the addition, multiplication on a_in, b_in

Strategy: Let us push the arithmetic resources at the output side and the selection logic at input side.

RTL Tweaks: The above-mentioned strategy can result into the better area and performance for the design. Try to tweak the RTL with reference to Table 11.4.

This results into the pushing of the common resources at the output side and improves the design performance. The RTL description is shown in Example 3.

Example 3 RTL with optimization strategies

```
//////////////////////////////////////////////////////////////////////////////

module area_RTL_optimization( input a_in,b_in, c_in, d_in, e_in, f_in,
                    input sel_in_0, sel_in_1,
                    output reg y1_out, y2_out,
                    output reg [1:0] y3_out, y4_out);

 reg tmp1,tmp2, tmp3,tmp4;

always @ *
begin
  if ( sel_in_0)
  begin
         tmp1 = a_in;
   end
   else
   begin
          tmp1 = c_in;

   end
 end

 always @ *
 begin
   if ( sel_in_0)
   begin
    tmp2 = b_in;
    end
    else
    begin
    tmp2 = d_in;
    end
  end

  always @ *
  begin
```

```
    y1_out = tmp1 + tmp2;
    y3_out = tmp1 * tmp2;
  end
always @ *
  begin
    if ( sel_in_1)
    begin
     tmp3 = e_in;
     end
     else
     begin
     tmp3 = a_in;
     end
   end

   always @ *
   begin
     if ( sel_in_1)
     begin
      tmp4 = f_in;
      end
      else
      begin
      tmp4 = b_in;
      end
    end

    always @ *
    begin

      y2_out = tmp3 + tmp4;
      y4_out = tmp3 * tmp4;
    end
endmodule

//////////////////////////////////////////////////////////////////////////////
```

As shown in the synthesis schematic, it improves area as the design has only two adders and two multipliers (Fig. 11.4).

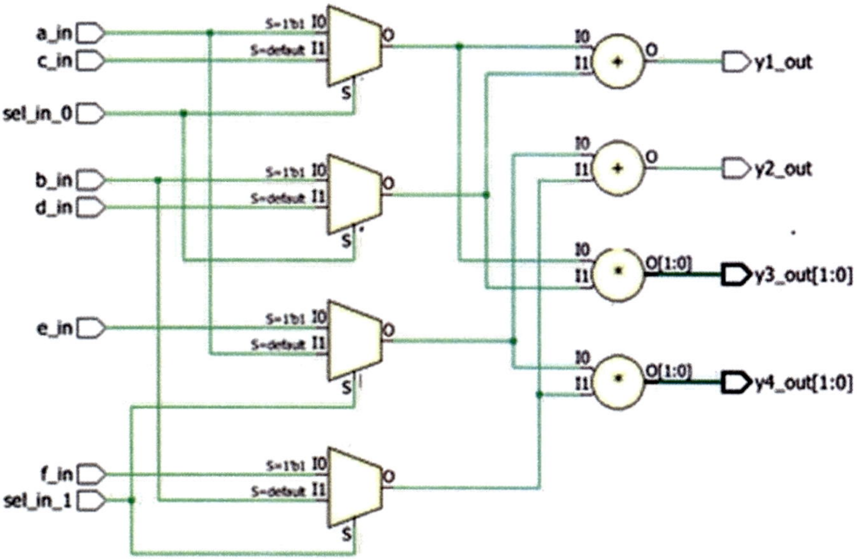

Fig. 11.4 RTL schematic for the design with area optimization

11.6 Synthesis Optimization Techniques Using RTL Tweaks

During the design synthesis, if the performance is not met then following can be few of the options

1. Perform the synthesis with the goal of optimization using tool-based commands. These techniques are discussed in the Chap. 12.
2. Perform the RTL tweaks with the goal to improve for the area, speed and then perform the synthesis.
3. Perform the architecture and micro-architecture tweaks and then try to tweak the RTL and run the synthesis.

The RTL tweaks which are useful to boost the design performance are discussed in this section.

11.6.1 Resource Allocation

This is used for the better synthesis results, and this optimization technique uses the sharing of hardware resources.

Consider the Verilog procedural block described in Example 4.

Example 4 RTL without any common resource allocation

```
//////////////////////////////////////////////////////////////////////

module resource_allocation( input a_in, b_in, c_in, d_in, sel_in,
output reg y_out);

always @ *
begin
 if (sel_in)
        y_out = a_in + b_in;

 else
 y_out = c_in + d_in;

 end
endmodule

//////////////////////////////////////////////////////////////////////
```

The RTL description (Example) infers two adders to perform the addition. It also infers the 2:1 MUX to select one of the adder outputs. The synthesis schematic is shown in Fig. 11.5.

As discussed in Sect. 11.5 if the common resources are pushed at the output side, then the logic inferred will have the single arithmetic resource (adder). The RTL tweak is described in Example 5.

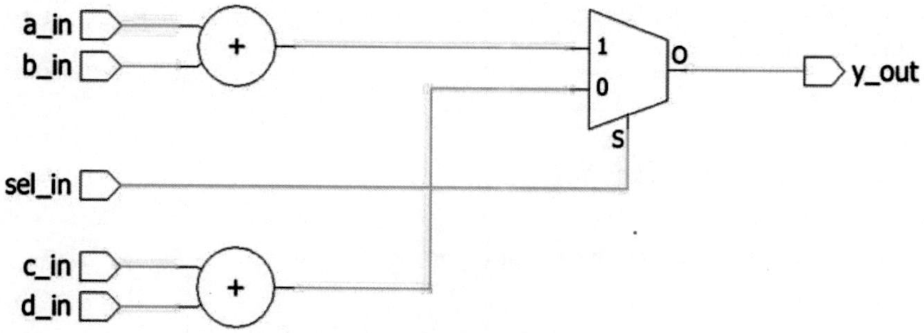

Fig. 11.5 Synthesis result without resource allocation

Example 5 RTL with common resource allocation

```
////////////////////////////////////////////////////////////////////////

module resource_allocation( input a_in, b_in, c_in, d_in, sel_in,
output y_out);

 reg tmp1, tmp2;
 always @ *
 begin
  if (sel_in)
  begin
          tmp1 = a_in ;
          tmp2 = b_in;
  end
  else
  begin
          tmp1 = c_in ;
          tmp2 = d_in;
  end
  end

  assign y_out = tmp1 + tmp2;
 endmodule

////////////////////////////////////////////////////////////////////////
```

This technique is called as resource sharing and useful to improve the area of the design (Fig. 11.6).

11.6.2 Dead Zone Elimination

The piece of the code which is never executed is called as dead zone code. The dead zone code elimination technique needs to be used for the better synthesis results.

The Verilog RTL is shown in Example 6. As described in the RTL $a = 4$ and $b = 3$ so the condition $a >$ b is always true and hence the some piece of code (check else clause) is always false and the y2_out = 1 so the synthesis tool will trim the

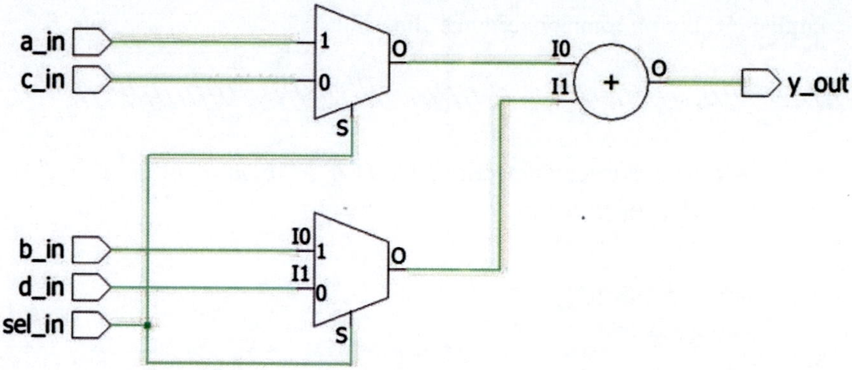

Fig. 11.6 Synthesis result with resource allocation

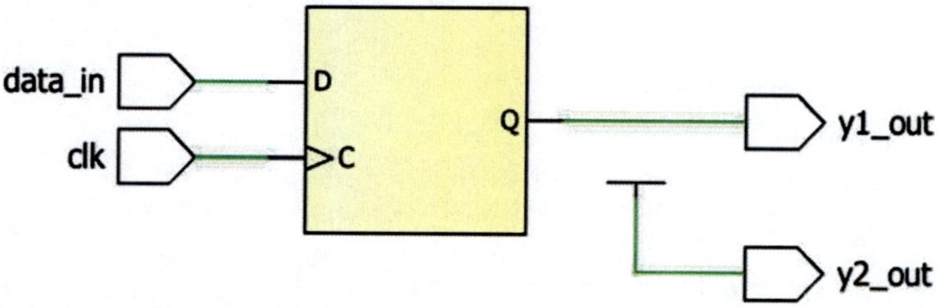

Fig. 11.7 Logic trimming due to dead zone-1

large multiplexers inferred due to if-else. So, the RTL tweaks are recommended. Use constant passing that is y2_out = 1 instead of if-else within the always procedural block (Fig. 11.7).

Example 6 RTL with dead zone-1

```
////////////////////////////////////////////////////////////////////////////

module dead_zone( input data_in, clk, output reg y1_out, y2_out );

integer a = 4;
integer b= 3;

always @ *
begin
 if (a>b)
          y2_out = 1;

 else
          y2_out = 0;
end
always @ (posedge clk)
begin
  y1_out <= data_in;
 end
 endmodule
////////////////////////////////////////////////////////////////////////////
```

The Verilog RTL is shown in Example 7. As described in the RTL $a = 3$ and $b = 4$ so the condition $a >$ b is always false and hence the some piece of code (check if clause) is always false and the y2_out = 0 so the synthesis tool will trim the large multiplexers inferred due to if-else. So, the RTL tweaks are recommended. Use constant passing that is y2_out = 0 instead of if-else within the always procedural block (Fig. 11.8).

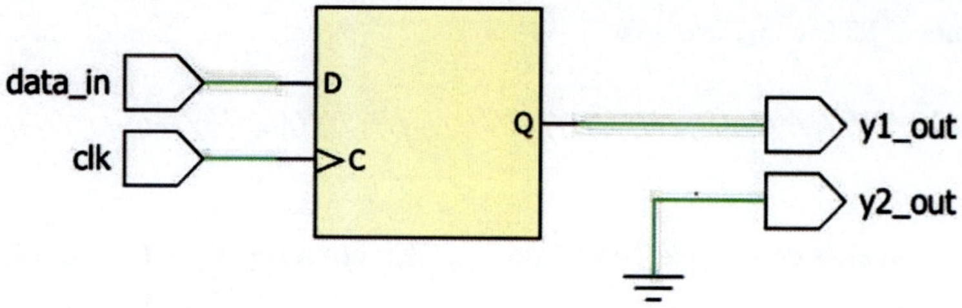

Fig. 11.8 Logic trimming due to dead zone-2

Example 7 RTL with dead zone-2

```
////////////////////////////////////////////////////////////////////////////////

module dead_zone ( input data_in, clk, output reg y1_out, y2_out
);

integer a = 3;
integer b= 4;

always @ *
begin
 if (a>b)
          y2_out = 1;

  else
         y2_out = 0;
end
always @ (posedge clk)
begin
         y1_out <= data_in;
 end
endmodule

////////////////////////////////////////////////////////////////////////////////
```

11.6.3 Use of Parentheses

The RTL team members should use the parenthesis and grouping of the terms to avoid the cascade logic. In the RTL description (Example 8), the synthesis tool infers the priority logic, where a_in has highest priority over other inputs and infers the design with the combinational logic which as the maximum delay.

Example 8 RTL without parenthesis

```
////////////////////////////////////////////////////////////////////////////////

module grouping_terms(input a_in, b_in, c_in, d_in, e_in,f_in,g_in,h_in,
output y_out );


assign y_out = a_in & b_in & c_in & d_in & e_in & f_in & g_in & h_in ;

endmodule

////////////////////////////////////////////////////////////////////////////////
```

If each stage has delay of 0.5 ns, then the overall delay is around 3.5 ns (Fig. 11.9).

Fig. 11.9 Cascade or priority logic

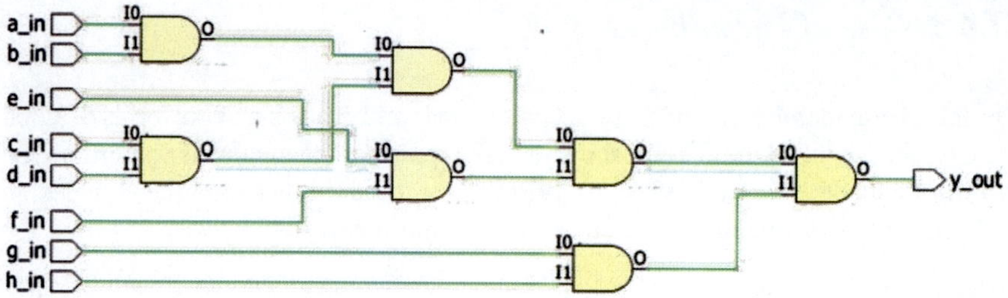

Fig. 11.10 Logic inferred after RTL tweak strategy-1

Strategy 1

The RTL (Example 8) is tweaked using the grouping of the terms within the parenthesis, and it infers the logic with cascade four stages. The delay of each stage is 0.5 ns; then, the overall combinational logic delay is 2 ns (Fig. 11.10).

Example 9 RTL with parenthesis-1

```
////////////////////////////////////////////////////////////////////////////////

module grouping_terms(input a_in, b_in, c_in, d_in, e_in,f_in,g_in,h_in,
output y_out );


assign y_out = (a_in & b_in) & (c_in & d_in) & (e_in & f_in) & (g_in &
h_in) ;

endmodule
////////////////////////////////////////////////////////////////////////////////
```

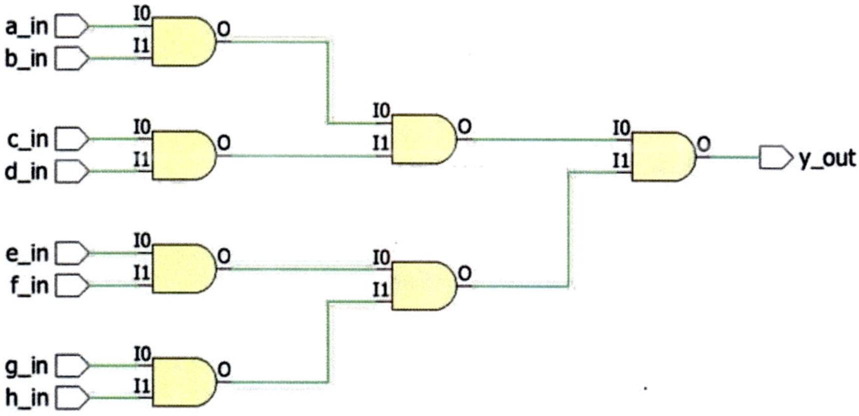

Fig. 11.11 Logic inferred after RTL tweak strategy-2

<u>**Strategy 2**</u>
Try to have the RTL tweak with better structuring and grouping of the terms to infer logic with the three stages. This can be multiplexed logic or parallel logic.

Example 10 RTL with parenthesis-2

```
/////////////////////////////////////////////////////////////////////////////

module grouping_terms(input a_in, b_in, c_in, d_in, e_in,f_in,g_in,h_in,
output y_out );


assign y_out = ((a_in & b_in) & (c_in & d_in)) & ((e_in & f_in) & (g_in &
h_in)) ;

endmodule
/////////////////////////////////////////////////////////////////////////////
```

Due to grouping, the logic has three stages and the overall delay is 1.5 ns, thus with reference to original RTL the team is successful to reduce the delay by 2 ns hence the improvement in the design performance (Fig. 11.11).

11.6.4 Grouping the Terms

Consider the RTL description shown in Example 11. During synthesis, the design infers the cascade logic using the arithmetic resources (adder, subtractor). If the

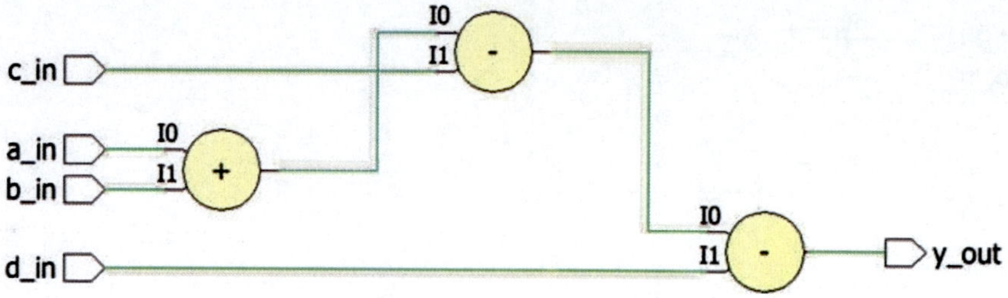

Fig. 11.12 Synthesis result without use of parentheses

propagation delay of the adder is 1 ns and subtractor is 0.5 ns, then overall delay is 2.0 ns (Fig. 11.12).

Example 11 RTL without grouping

```
////////////////////////////////////////////////////////////////////////////////

module grouping_terms(input a_in,b_in,c_in,d_in, output y_out);

assign y_out = a_in + b_in - c_in -d_in;

endmodule

////////////////////////////////////////////////////////////////////////////////
```

If the RTL described in Example 12 is tweaked using the parenthesis to eliminate the cascade stage, then the design performance can be improved by few nanoseconds. Check for the RTL tweak recommended in the Example.

Example 12 RTL with grouping

```
////////////////////////////////////////////////////////////////////////////////

module grouping_terms(input a_in,b_in,c_in,d_in, output y_out);

assign y_out = (a_in + b_in) – (c_in +d_in);

endmodule

////////////////////////////////////////////////////////////////////////////////
```

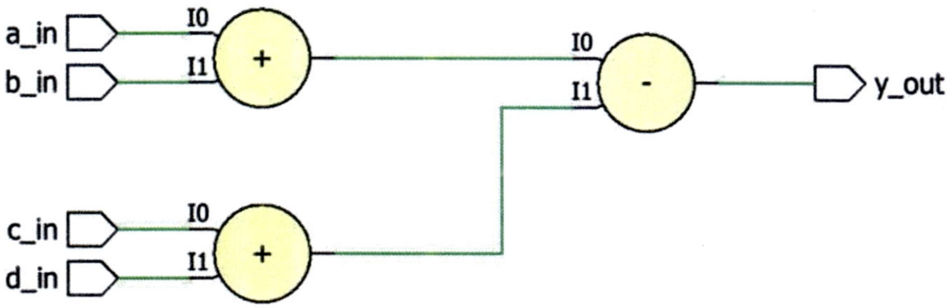

Fig. 11.13 Synthesis result with use of parentheses

The synthesis tool during optimization phase will infer the two-stage logic, and the overall delay is 1.5 ns thus reduction of 0.5 ns delay (Fig. 11.13).

11.7 FPGA Synthesis

It uses the dedicated FPGA resources to represent the gate-level netlist. These resources are CLB (slice register and LUTs) IOBs (Figs. 11.14 and 11.15).

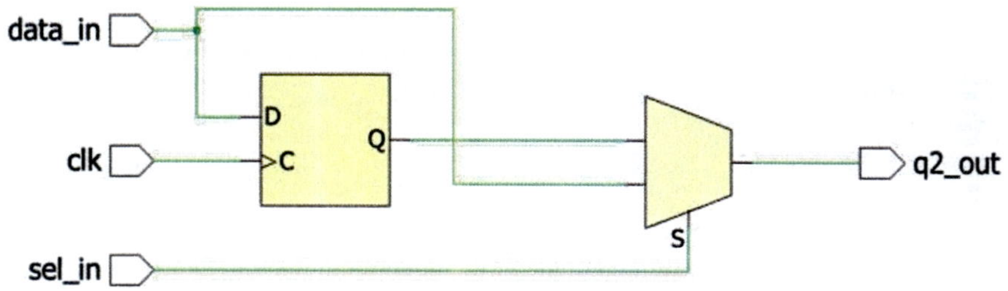

Fig. 11.14 FPGA logic inferred using the slice register and MUX

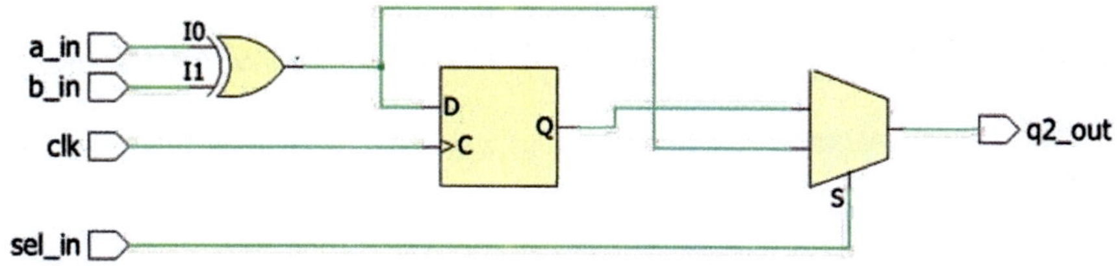

Fig. 11.15 FPGA logic inferred using CLB (LUT, register, MUX)

Example 13 Sequential logic description

```
////////////////////////////////////////////////////////////////////////////////////////


module fpga_design (input clk, data_in, sel_in, output  q2_out);
reg q1_out;
  always @ (posedge clk)
  begin
          q1_out <= data_in;
  end

  assign q2_out = (sel_in) ? q1_out : data_in;
endmodule

////////////////////////////////////////////////////////////////////////////////////////
```

Example 14 Sequential logic description-1

```
////////////////////////////////////////////////////////////////////////////////////////

module fpga_design(input clk, a_in,b_in,sel_in, output  q2_out);
reg q1_out;
  always @ (posedge clk)
  begin
          q1_out <= q_out;
  end
  assign q_out = a_in ^ b_in;
  assign q2_out = (sel_in) ? q1_out : q_out;
endmodule

////////////////////////////////////////////////////////////////////////////////////////
```

ASIC and FPGA synthesis differ in various ways, and for more details, refer the Chaps. 18 and 19.

11.8 Chapter Summary

Following are important points to conclude the chapter

1. Synthesis tool carries out the optimization to meet the area and speed constraints.
2. Optimization is based on various cost functions.
3. RTL tweaks such as resource sharing, dead zone elimination and the grouping of terms are useful to improve the design performance.
4. FPGA synthesis tool uses the FPGA resources such as LUTs, IOBs, slice registers to infer the logic.
5. The cascade stages will have more delay as compared to parallel logic.
6. Use the multiplexed logic to improve the design performance.

Chapter 12
Synthesis and Optimization Techniques

As discussed in the previous chapters, the synthesis is carried out to get the lower level of abstraction of the design. To have the gate-level netlist, we will perform logic synthesis, and to have the device or switch-level abstraction, we will perform the physical synthesis. The chapter is useful to understand the ASIC synthesis using Synopsys DC and the optimization techniques used to meet the desired constraints. The synthesis for the complex design is performed at the block and top level, and during the logic synthesis, our goal is to meet for the area and speed constraints. Synopsys DC is not used to optimize for the power. The next subsequent sections discuss the synthesis and optimization techniques.

12.1 Introduction

Use the ***synopsys_dc.setup*** *to setup the design compiler for the synthesis and optimization.* There should be two startup files; one should be in the current working directory, and another should be in the root directory where the Design Compiler is installed. To use the tool, the following important parameters need to be setup.

1. **search_path**: This parameter is used to search for the synthesis technology library and used as the reference during synthesis.
2. **target_library**: This parameter is used by the synthesis tool while mapping the logic cells. The target library contains the logic cells.
3. **symbol_library**: All the logic cells have symbolical representation. The parameter is used to point to the library which has the visual information for the logic cells present in the technology synthesis library.

Tool-based optimization strategies using Synopsys DC are useful to improve the performance of the design.

V. Taraate, *ASIC Design and Synthesis*,
https://doi.org/10.1007/978-981-33-4642-0_12

Table 12.1 Design objects used by synthesis tool

Design object	Description
Cell	Cell is also called as instance. The instantiated name of the sub-design is called as cell
Reference	It is original design to which cell or instance refers. For example, instantiated sub-design must refer to the design which consists of the functional description of the sub-design
Ports	The primary inputs and outputs or IO's of the design are called as ports
Pins	The primary inputs, outputs, IO's of cells in the design are called as pins
Net	Wires used for the connection between ports of the pins of the different designs are called as net
Clock	The input port or pin used as clock source is called as clock
Library	The technology-specific cells used for targeting for synthesis, linking or for reference are called as library

4. **link library**: The tool uses the cells from the target_library during mapping the functionality; this parameter is used to point to the logic gates in the synthesis technology library.

The above four parameters for **.synopsys_dc.setup** are specified by using following

```
set search_path "./synopsys/libraries/syn/cell_library/syn"
set target_library "tcbn65lpwc.db, tcbn65lpbc.db"
set link_library "$target_library $symbol_library"
set symbol_library "standard.sldb dw_foundation.sldb"
```

Once the above parameters are set up for the required library, then the synthesis tool can be invoked at the command prompt.

Every design is the description of the logic circuit to perform some of the operations. The design can be single module description or can consist of the multiple modules. The design objects are described in Table 12.1.

12.2 Synthesis Using Design Compiler

The synthesis tool uses the RTL design Verilog (.v) files, constraints (.sdc) and library (.lib) as an inputs to get the optimized gate-level netlist using standard cells available in the library. During the ASIC synthesis, few steps are performed, and these are mainly translate, map and optimize. Figure 12.1 gives the brief information about the ASIC synthesis steps to generate the gate-level netlist.

1. **Read Library**: To perform the logic synthesis, the synthesis tool reads the DesignWare libraries, technology libraries, and symbol libraries.
   What the DesignWare library consists of?
   The DesignWare library consists of the complex cells such as adders, comparators, multipliers, etc.
   What the technology library consists of?
   The technology library consists of the logic gates, flip-flops, and latches.
   During synthesis, the synthesis tool algorithms determine when to use the technology library cells and when to use the DesignWare library components. These library cells are used efficiently to generate the gate-level netlist.

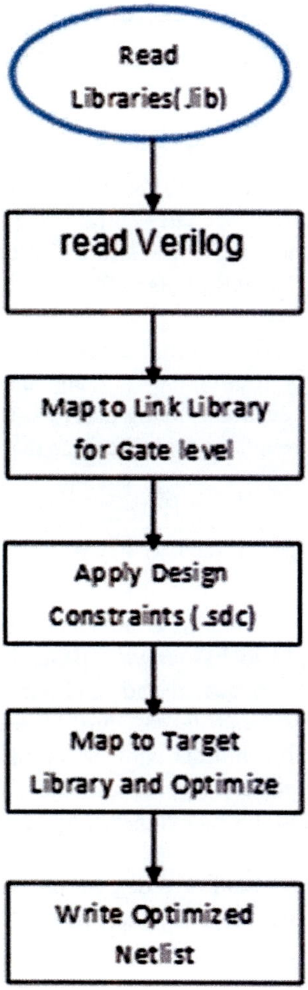

Fig. 12.1 ASIC synthesis steps

2. **Read RTL Description**: The next step is to read the RTL description that is Verilog (.v) source files.
3. **Map the link Library**: The synthesis tool after reading the libraries and the RTL description performs few important steps.
   (a) Design optimization.
   (b) Technology-independent optimization.
   (c) Mapping the logic using the technology library. The above process is called as linking the logic to the desired target library. So basically, link library can be IO library, cell library or macro library and used to link the design, and target library is used while optimizing the design.
4. **Use Design Constraints**: The synthesis tool uses the design constraints such as area, speed and power while optimizing the design using the standard cells which are available in the target library. The DC doesn't optimize for the power and power planning and optimization we can have during the physical design flow.
5. **Map the Design to Target Library**: For efficient RTL coding, it is required that RTL design engineer should have good understanding of the target standard cell library. After the design is optimized, then the design is ready for the Design for Testability (DFT) that is to detect early faults in the design. During RTL design stage only, the DFT-friendly RTL needs to be described to enable quick scan insertions and testing for various faults in the design.
6. **Optimize and Save Netlist**: The optimized netlist can be in the Verilog (.v) format or in the database (.ddc) format and will be used by the placement and routing tool. Based on the routing the back-annotation can be performed with actual routing delays for accurate timing analysis. If timing goals are not met, then the design can go through the resynthesis so that the timing goals can meet.

12.3 Synthesis and Optimization Flow

Modern ASIC designs are extraordinarily complex in the and consists of few million or billion gates. Design complexity has grown exponentially in the past few decades due to the demand of the sophisticated and intelligent devices and IPs. In such scenario, there is additional overhead and cost during the design synthesis and timing closure. In such scenario, the synthesis with the optimization goal can be better technique to meet the block- and top-level constraints. The Synopsys DC is leading EDA tool used to perform the design synthesis, and Synopsys PT is used for the timing closure.

The design constraints are classified as design rule constraints and optimization constraints and discussed in the Chaps. 10 and 11.

The synthesis and optimization flow is shown in Fig. 12.2. These are also treated as the steps while carrying out synthesis for any design. The compilation strategy can be chosen as top-down or bottom-up. The commands used during synthesis are discussed in the subsequent session.

1. **Read Design Object**: Design object is Verilog RTL code which is simulated for the functional correctness. The commands used at this step are

   analyze, elaborate, read

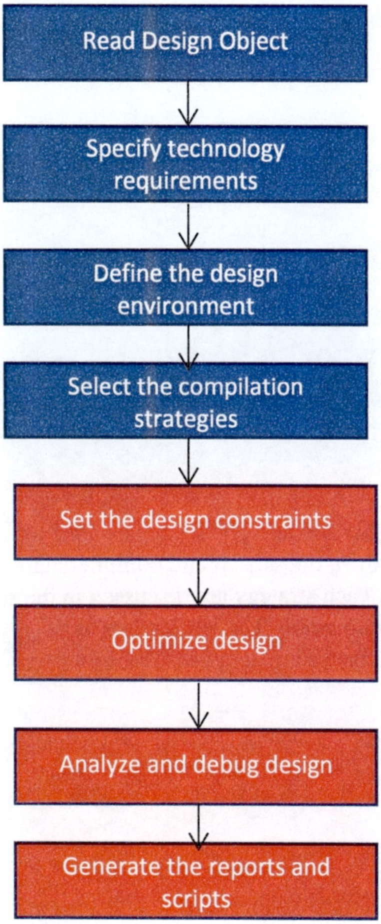

Fig. 12.2 Flow for synthesis and optimization

2. **Specify Technology Requirements**: In these steps, the design rules and libraries required need to be specified. The commands used in this step are

Library Objects
link_library
target_library
symbol_library
Design Rules
set_max_transition
set_min_transition
set_max_fanout
set_min_fanout
set_max_capacitance
set_min_capacitance

3. **Define the Design Environment**: The design environment includes the process, temperature, voltage conditions, drive strength and effect of load driving the design. The commands used are

set_operating_conditions
set_wire_load
set_drive
set_driving_cell
set_load
set_fanout_load

4. **Select Compilation Strategies**: The strategies used for optimizing hierarchical design include top-down, bottom-up and compile-characterize. The advantages and disadvantages of each strategy are discussed in the subsequent section.
5. **Set the Design Constraints**: The constraints need to be set for the design optimization and for the timing analysis. The commands used in this step are

create_clock
set_clock_skew
set_input_delay
set_output_delay
set_max_area

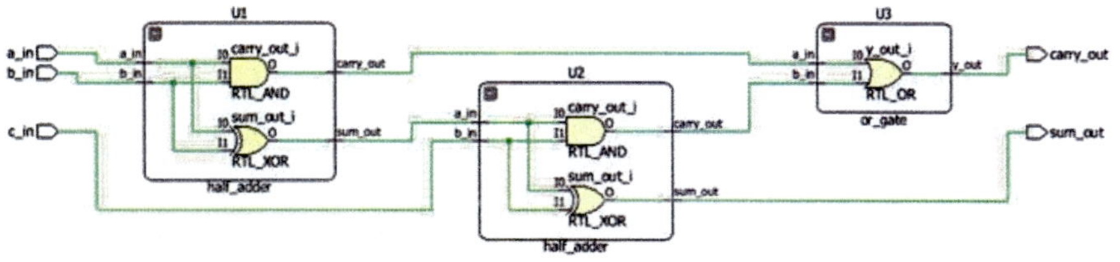

Fig. 12.3 Full adder schematic

6. **Optimize Design**: Perform the design synthesis to generate technology-specific gate-level netlist. The command used is

```
compile
```

7. **Analyze and Debug the Design**: This step is important to understand the potential issues in the design by generating various reports. The commands used in this step are

```
check_design
report_area
report_constraint
report_timing
```

8. **Generate Various Reports and Scripts**: The design database is stored in the form of script file.

Consider the top-level object as full adder with inputs 'a_in, b_in, c_in' and outputs 'sum_out, carry_out' (Fig. 12.3).

The top-down compilation run is shown by using the following script and can be used in the practical scenario. To synthesize the design and to compile, use the script shown in Example 1.

Example 1 Key steps for synthesis and compilation

```
/* read the design object */
read -format verilog full_adder.v
/* specify the technology requirements */
target_library = my_library.db
symbol_library = my_library.sdb
link_library = "*" + target_library
/* define the design environment */
set_load 2.0 sum_out
set_load 1.2 carry_out
set_driving_cell -cell FD1 all_inputs()
set_drive 0 clk_name
/* set the design constraints */
set_input_delay 1.25 -clock clk {a_in, b_in}
set_input_delay 3.0 -clock clk c_in
set_max_area 0
/* synthesize the design */
compile
/* generates reports */
report_constraint
report_area
/* save the design database */
write -format db -hierarchy -output full_adder.db
```

12.4 Area Optimization Techniques

There are several techniques used for minimizing the overall area of the design. The highest priority of the designer is to optimize for the timing followed by area. There are several efficient area minimization techniques at the RTL level. In the previous section, we have discussed the resource sharing. Following are the key guidelines used to optimize for the area

1. Avoid use of the combinational logic as individual block or module
2. Do not use the glue logic between two modules
3. Use ***set_max_area*** attribute while synthesizing the design.

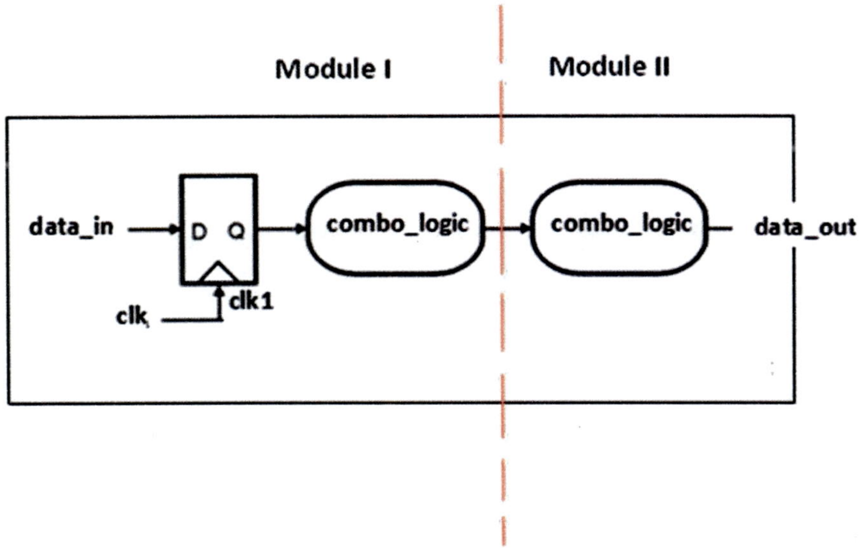

Fig. 12.4 Combinational logic as individual module

12.4.1 Avoid Use of Combinational Logic as Individual Block

It is recommended that do not use the combinational logic as individual block. If the individual combinational module is used, then design compiler will not be able to optimize the individual block. This is not a good design partitioning. The hierarchy of the module is fixed, and design compiler will not be able to optimize for the hierarchical combinational designs. Consider the scenario shown in Fig. 12.4. It has module I and module II, module II is separate combinational block so the design compiler will not be able to optimize module II, as design compiler doesn't optimize the port interfaces.

If the design is partitioned properly, then the overall optimization will boost the design performance. A better partitioned ASIC design should have combined functionality of module I and module II. The functionality of $A + B$ in the single module is shown in Fig. 12.5 and results into the faster optimization for the design.

12.4.2 Avoid Use of Glue Logic Between Two Modules

The glue logic between two different blocks is shown in Fig. 12.6. Such type of design partitioning strategy is not good and will insert more combinational delays, the reason being the logic gate cannot be optimized by the design compiler. To avoid this type of scenario, it is recommended to use the ***group*** command. Either group

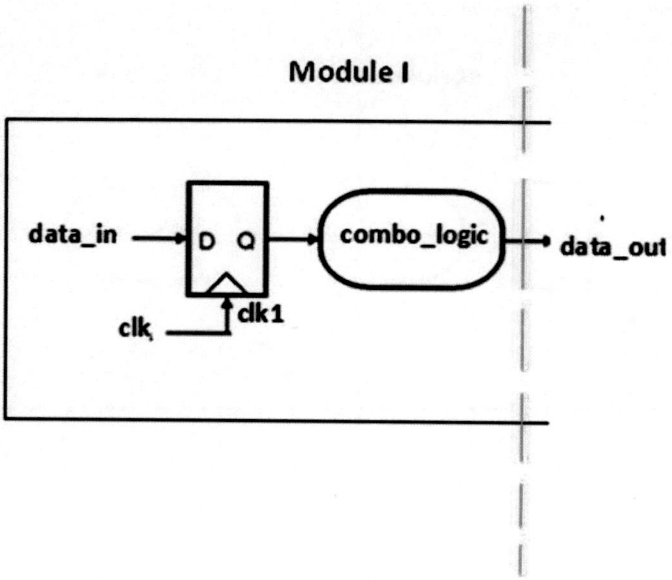

Fig. 12.5 Eliminating individual combinational module

the glue logic in the module I or module II. Following command used to group the glue logic into module I

dc_shell> group {m1, m3} –design_name moduleIII cell_name or_gate

Following command used to group the glue logic into module II

dc_shell> group {m2, m2} –design_name moduleIII cell_name or_gate

12.4.3 Use of set_max_area Attribute

To optimize for the area, it is recommended to use the attribute ***set_max_area***. This attribute is effective in the optimization of the design. Design compiler gives the highest priority to the timing optimization. If timing is met, then only the area optimization phase can start. The priorities for the design optimization are listed below

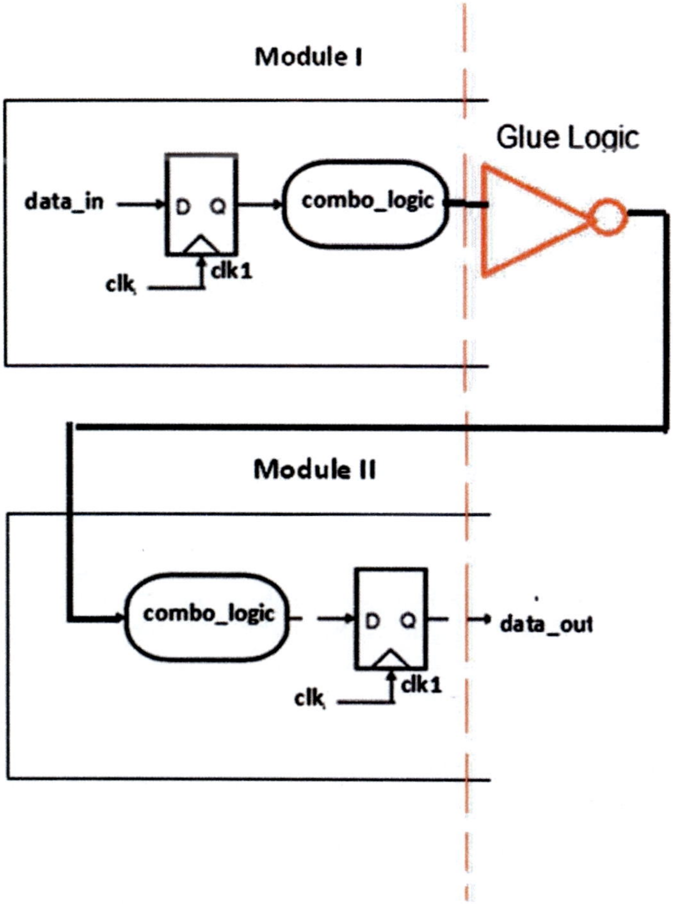

Fig. 12.6 Glue logic between two blocks

1. Design rule constraints (DRC)
2. Timing
3. Power
4. Area.

12.4.4 Area Report

The area is reported using ***report_area*** command. The sample area report is shown the Example 2. The area report for any design consists of the number of ports, nets, references. It also gives information about the combinational, sequential, and total cell area.

Example 2 Area report

Number of ports:	3
Number of nets:	8
Number of cells:	7
Number of references:	2
Combinational area:	100.349998
Non combinational area:	125.440002
Net Interconnect area: undefined (Wire load has zero net area)	
Total cell area:	225.790009
Total area:	undefined

12.5 Design Partitioning and Structuring

The design needs to be partitioned for the better synthesis and optimization. Partitioning is carried at functional level by considering the interface boundaries and the clock and power domains. It is the practical reality that the design which is better partitioned generates better synthesis results and even it reduces the synthesis runtime. The following are important guidelines recommended for the design partitioning

1. Partition the design for the design reuse.
2. For the different functionality, use the different module. That is use the modular approach during the design.
3. Use the combinational logic in the same block which reduces the insertion delays.
4. Use the separate block or structure logic for the random logic.
5. Partition the design at the top level.
6. Do not use the glue logic at the top level.
7. Use the separate module for state machines that is isolating the state machines form the other logic.
8. Limit the logic size to maximum 10K gates for every block.
9. Avoid use of the multiple clocks in the same block.
10. Isolate the synchronizers for the multiple clock domain designs (Fig. 12.7).

Consider the design which already we have discussed during the Chap. 9.

Clock domain 1: It is controlled by the clk1, and the functional blocks of this clock domain are

1. ALU
2. Internal memory
3. Interrupt controller
4. Pointers and counters

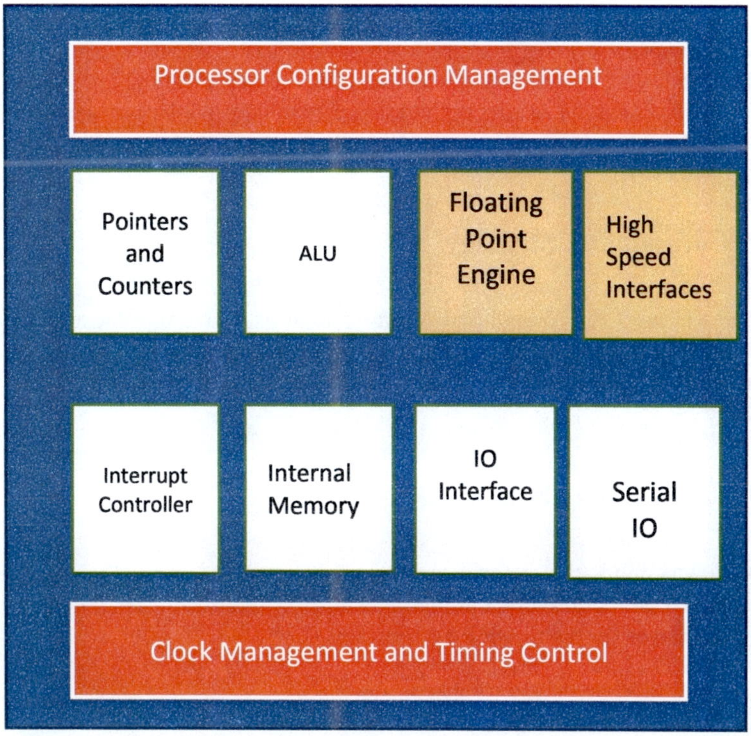

Fig. 12.7 Design with multiple clock domains and functional blocks

5. Serial IO
6. IO interfaces.

In the architecture clock domain 1 block is indicated by the yellow color.

Clock domain 2: It is controlled by the clk2, and the functional blocks of this clock domain are

1. Floating Point Unit
2. High-speed interfaces
   In the architecture clock domain 2 blocks are indicated by the white color.
   We can have the strategy for the synthesis and important highlights are
   (a) Have clock and constraint definitions for each clock domain.
   (b) Perform the block-level synthesis using block-level constraints.
   (c) Perform the top-level synthesis using the top-level constraints.
   (d) Use the strategies as bottom-up of top-down during the compilation. It depends on the design complexity.

(e) Use do not touch attribute if the constraints are met during the block-level synthesis.
(f) Check for the violators and fix them run the optimization.
(g) Recompile with various options and perform the synthesis to lead the better design optimization.

12.6 Compilation Strategy

The methods used for compilation of any design can have top-down or bottom-up compilation approach. Each compilation method has its own advantages and disadvantages.

12.6.1 Top-Down Compilation

The top-down compilation uses the top-level design constraints and easier to execute as compare to the bottom-up compilation approach. Following are the advantages and disadvantages for the top-down compilation

Advantages

1. Optimization engines work on full design, complete paths
2. Usually get best optimization result
3. No iteration required
4. Simpler constraints
5. Simpler data management.

Disadvantages

1. Longer runtime
2. More memory requirements
3. More runtime.

The commands used for the top-down compilation are

```
dc_shell> current_design TOP
dc_shell> compile –timing_high_effort_script
```

12.6.2 Bottom-Up Compilation

The bottom-up compilation compiles submodule first, and then, it moves toward top level. The care must be taken by the designer to set "***set_dont_touch***" attribute on the submodules to avoid recompilation of the submodules. The designer needs to know the timing information for the inputs and outputs for each of the submodule. The advantages and disadvantages are discussed below

Advantages

1. Faster as compared to top-down compilation
2. Less processing required per run
3. Less memory requirement.

Disadvantages

1. Optimization works on the submodule or subdesign
2. More iteration required
3. More hierarchies to maintain.

Consider the design has two submodules. The commands used for the bottom-up compilation are

```
dc_shell> current_design submodule1
dc_shell> compile –timing_high_effort_script
dc_shell> set_dont_touch submodule1
dc_shell> current_design submodule2
dc_shell > compile –timing_high_effort_script
dc_shell> set_dont_touch submodule2
dc_shell> current_design TOP
dc_shell> compile –timing_high_effort_script
```

12.7 Chapter Summary

Following are the important points to conclude the chapter

1. Constraints are of mainly optimization and design rule constraints.
2. Do not partition the design across combinational boundaries as it incurs significant amount of insertion delay.

3. Choose for the compilation strategies either top-down or bottom-up depending on the design complexity.
4. Synopsys DC will not be able to optimize for the power.
5. Partition the design by considering the various clock domains and power domains.
6. Use the recompile with option to meet the optimization goals.

Chapter 13
Design Optimization and Scenarios

During the logic optimization, we will try to optimize for the area and speed. The following sections are useful to address few of the optimization strategies used for during the ASIC synthesis. These strategies can be used during synthesis and optimization of the complex ASIC designs.

With the optimization constraints and performance improvement, the design rule constraints are discussed in this chapter. The priorities for the design optimization are listed below

1. Design rule constraints (DRC)
2. Timing
3. Power
4. Area.

13.1 Design Rule Constraints (DRC)

The main important design rule constraints are fanout, capacitance, and transition. These constraints are having high priority during the synthesis as compared to optimization constraints.

Design optimization for the area and speed is important to have better performance of ASIC.

V. Taraate, *ASIC Design and Synthesis*,
https://doi.org/10.1007/978-981-33-4642-0_13

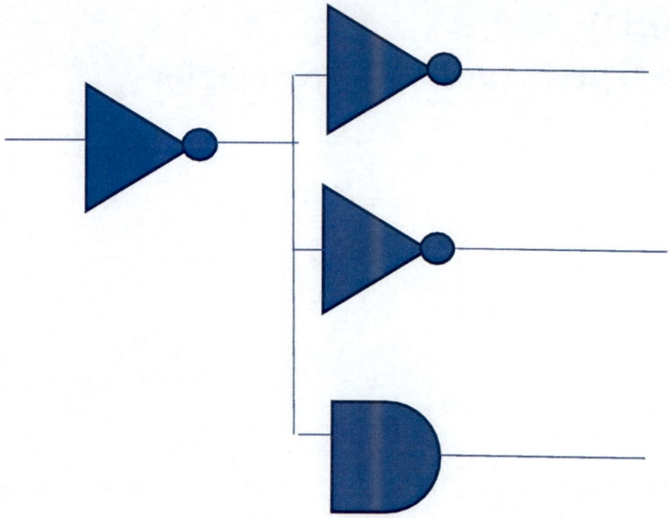

Fig. 13.1 Fanout of the cell

13.1.1 max_fanout

It is used to measure the number of loads a port or load can drive.

Consider the logic gate driving the multiple gates as shown in Fig. 13.1.

The technology library has information about the default fanout. Designer can use the following command to get the fanout

get_attribute library_name default_fanout_load

13.1.2 max_transition

The maximum transition from '0' to '1' or from '1' to '0' for specific net or entire design can be specified using Synopsys DC. We know the transition time is due to the RC time constant (Fig. 13.2).

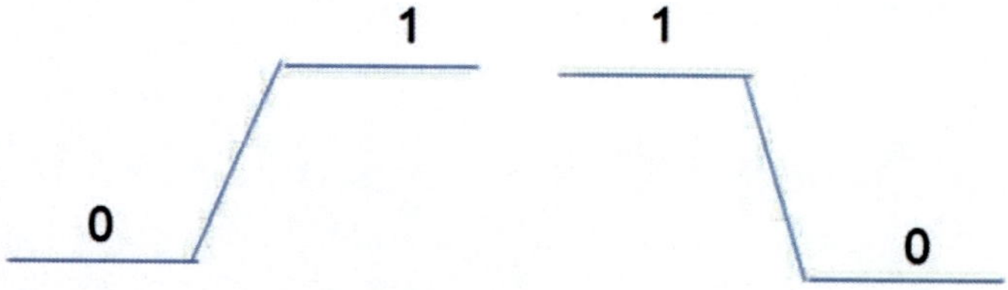

Fig. 13.2 Transition from '1' to '0', and '0' to '1'

How DC meets the specified maximum transition?

Consider the library which specifies maximum transition of '4' and designer specifies max_transition of '2'. What DC does is that it will try to meet the max_transition of 2.

The following Synopsys DC command is used to specify the max_transition.

set_max_transition <value> <design_name/port_name>

13.1.3 max_capacitance

It is used to provide information about the maximum net capacitance.

During the compilation, the DC takes care about the violations due to max_capacitance (Fig. 13.3).

So we know the max_capacitance constraint at output of driving cell is due to the net capacitance and the capacitance of pins driven by cell. Designer can use the following command to define max_capacitance

set_max_capacitance <value> <port_name/design_name>

As discussed, use the following commands to define the design rule constraints

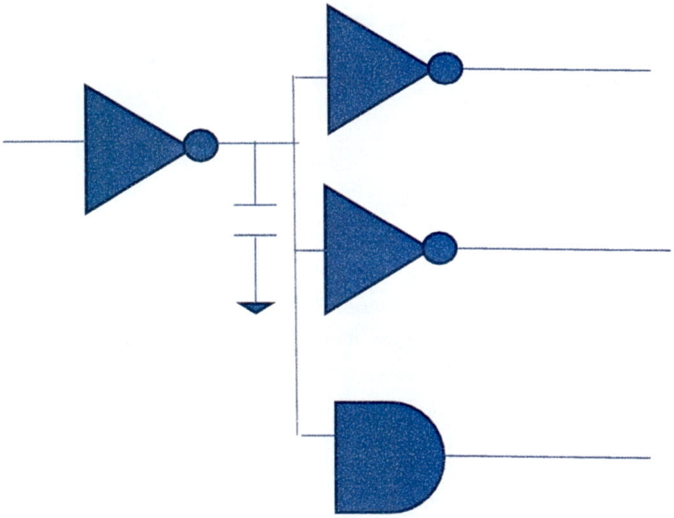

Fig. 13.3 Capacitance at driving cell

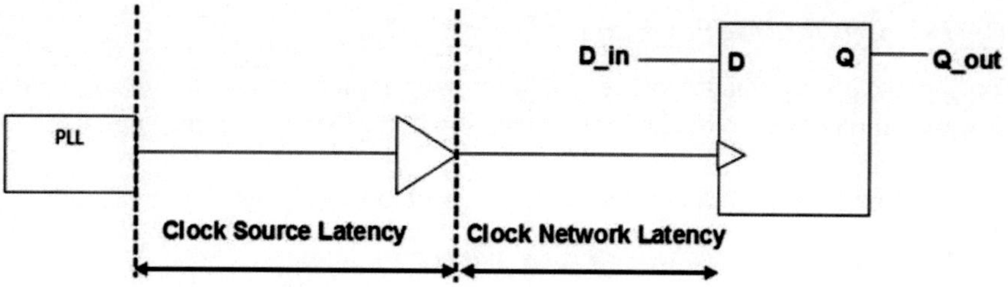

Fig. 13.4 Clock network latency

```
set_max_transition <value> <design_name/port_name >
set_max_fanout <value> <port_name/design_name >
set_max_capacitance <value> <design_name/port_name>
```

13.2 Clock Definitions and Latency

The clock definitions and latencies play an important role during the design. During the logic design, the clocking information that is information about the clock tree is not available. The section discusses various terms which we need to specify during the design synthesis.

13.2.1 Clock Network Latency

If we consider any ASIC, then the clock network latency and clock distribution decide the performance of any synchronous design. The PLL is used as clock source, and during STA it is essential to define the clock source and clock network latency. Figure 13.4 shows both the latencies.

13.2.2 Generated Clock

The generated clocks in the ASIC or SOC can be used as clocking source to the sequential blocks. The clocks are generated by using the clock divider networks. Figure 13.5 shows the generated clock using the clock divider. The useful Synopsys PT commands are described in Table 13.1.

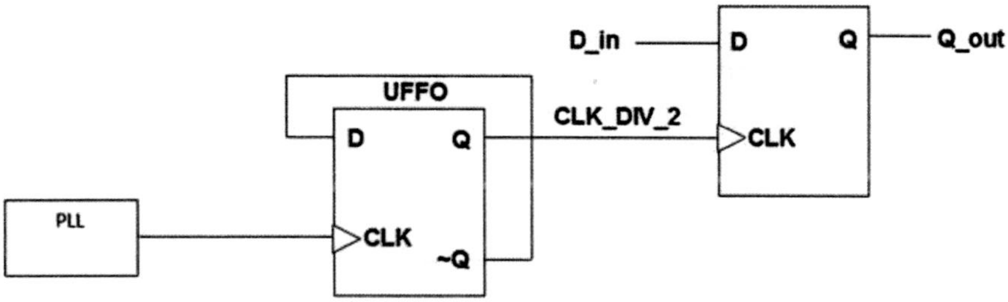

Fig. 13.5 Generated clock

Table 13.1 Clock and generated clock commands

Command	Description
create_clock –period 10 waveform {0 5} [get_ports clk_PLL]	Used to define clock having period 10 ns. The rising edge at 0 ns and falling edge at 5 ns
Create_generated_clock –name CLK_DIV_2 –source UPLL0/clkout –divide_by 2 [get_pins UFF0/Q]	Generated clock CLK_DIV_2 at q

Table 13.2 Commands to specify false path

Command	Description
set_false_path –from [get_clock Tclk_max] –to [get_clocks Tclk_min]	Used to set the false path between the Tclk_max and Tclk_min
set_false_path –through [get_pins UMUX/clk_select]	To set the false path with respect to clk_select

13.2.3 Clock Muxing and False Paths

Most of the times, we need to have clock multiplexing. The minimum and maximum clocks can be used in the design depending on the design requirements. During the ASIC testing, the minimum clock can be used. The false paths between these clocks need to be reported to the timing analyzer. To set the false path, use the command shown in Table 13.2 (Fig. 13.6).

13.2.4 Clock Gating

The clock gating checks need to be performed by the timing analyzer, and the command is described in Table 13.3 (Fig. 13.7).

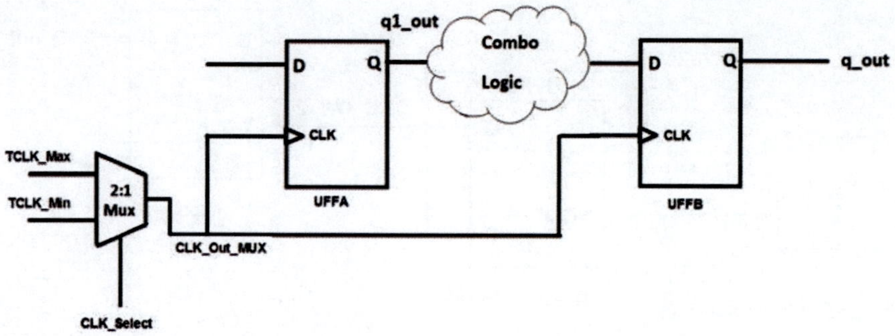

Fig. 13.6 Clock muxing and false path

Table 13.3 Clock gating checks commands

Command	Description
create_clock –period 10 [get_ports System_CLK]	To create the system clock of period 10 ns
create_generated_clock –name –divide_by 1 System_CLK [get pins UAND1/Z]	To get the same clock CLK_gate

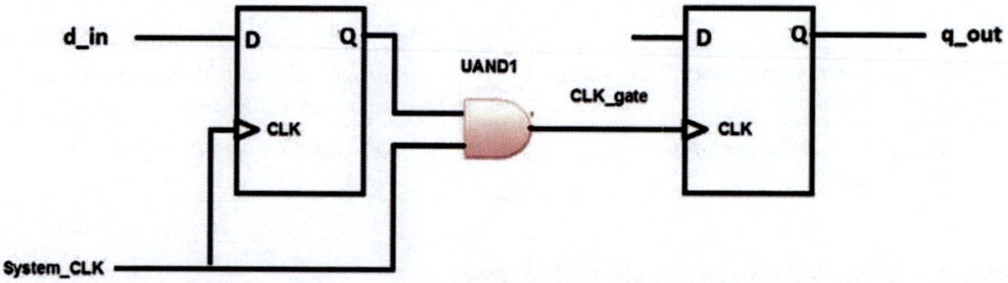

Fig. 13.7 Clock gating

13.3 Commands Useful During Design Synthesis and Optimization

The following section discusses the DC commands used during the optimization and performance improvement of the design.

13.3.1 set_dont_use

The command 'set_dont_use' can be used if the synthesis engineer wishes not to refer the cells from the technology library.

If 'don't_touch' attribute is used, then this command ignores the cells from the technology library.

The command is described below

set_dont_use library_name/cell_name

Consider cell name as XOR2; then, during the optimization the XOR2 cell is not used.

13.3.2 set_dont_touch

Most of the time during the optimization phase, we don't need to optimize the design which has already met the timing and area constraints. For example, consider the design shown in Fig. 13.8.

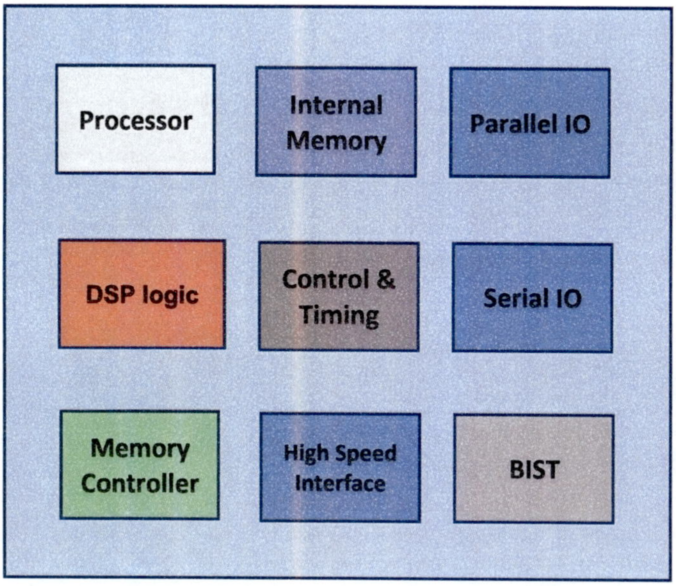

Fig. 13.8 SOC top-level design diagram

If the functional design processor, DSP logic, and memory controller have met the timing and area constraints and as further optimization is not required, the dont_touch command can be used.

The command is described below

set_dont_touch design_name

This we can prevent the synthesis re-optimization of the design functionality.

During logic synthesis, we need to use the hand-instantiated clock trees and as DC doesn't perform the synthesis for the clock tree use the don't_touch attribute for the same.

So if current design is soc_top, then we can do the following to allow don't touch of the functional block

current_design = soc_top
set_dont_touch u1
set_dont_touch u2
set_dont_touch u3

where the u1, u2, u3 are the instances of the processor, DSP logic, memory controller, respectively.

13.3.3 set_prefer

This command is used when the synthesis engineer wishes to change the priority of cells chosen by the DC during technology translation.

Now consider that the design netlist needs to be mapped to another technology library and then the command can be used and is described below

set_prefer library_name/libaray_name

13.3.4 Command for the Design Flattening

We can have the hierarchical or flattened design. Now consider the following expressions used in Verilog design.

Y1 = a_in & b_in;
Y2 = c_in & Y1;

The synthesis tool will infer the cascade logic and is shown in Fig. 13.9.

After flattening, we can get the expression as

Y2 = (c_in & a_in) & (c_in & b_in);

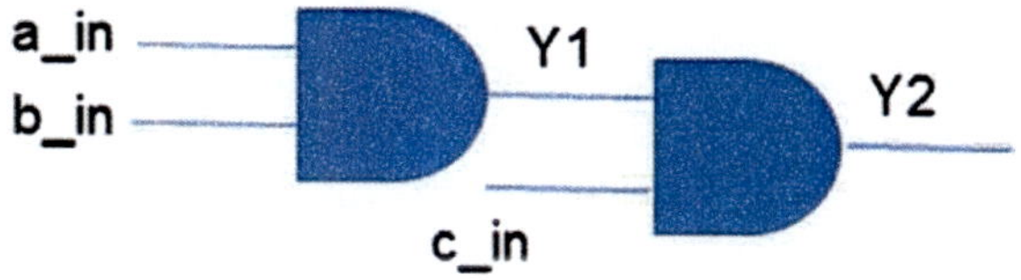

Fig. 13.9 Gate-level structure without flattening

set_flatten command is used to flatten the design.

The command is described as or may be referred as the switch

set_flatten true

If we refer the basic digital circuits, then the combinational logic is expressed using sum of product (sop). The SOP expression will improve the parallelism and the speed of the design.

Now let us imagine that can I flatten the multiplexers, XOR gates, and adders?

The answer is, it is not recommended to use flattening for these types of logic due to the use of the control elements. During optimization, the DC will never perform flattening unless and until it is specified.

Still there is always limitation to get the flattened netlist. It is recommended that if the design has 10 or less number of inputs then the flatten switch can be used.

13.3.5 Commands Used for Structuring

If the objective is to improve the area or gate count, then structuring is recommended.

The command is described below

set_structure –timing true

We can have the Boolean structuring or timing-driven structuring. By default, the DC uses the timing-driven structuring. To execute the Boolean structuring, we need to direct the synthesis tool.

13.3.6 Group and Ungroup Commands

To remove the hierarchy, the 'ungroup' command is used and to create the hierarchy the 'group' command is used.

The command is described below

ungroup –flatten –all

This will allow ungrouping of all the levels below soc_top design. But DC will take care that the design functionality for which the set_dont_touch attribute was applied will not be disturbed.

To ungroup all the synthetic designs before compiling, the following command can be used

replace_synthetic –ungroup

To group command can be used to have the new hierarchy. The command is give below

Group (u1, u2) –design_name name_block –cell_name soc_top_inst

So as discussed, the grouping forms the new instance soc_top_inst using the instances u1, u2.

13.4 Timing Optimization and Performance Improvement

While optimization, the timing has highest priority as compared to the power and area. During the first phase of optimization, the design compiler checks for the design rule constraints (DRC) violations, then the timing violations and the power constraints, and finally the area constraints. This section discusses the few timing optimization commands supported by the design compiler.

13.4.1 Design Compilation with 'map_effort high'

Most of the time, design engineer uses the option as ***map_effort medium*** while performing the synthesis. It is advisable that during synthesis of the first phase designer can use the option as map_effort medium as it reduces the compilation time. If the deign constraints are not met, then the designer can go for the incremental compilation with the option as ***map_effort high***. This can improve the design performance by at least 5–10%.

The sdc command is shown below

```
dc_shell> compile –map_effort_high –incremental_mapping
```

13.4.2 Logical Flattening

The design hierarchy of the design can be broken by using logical flattening of the design. The option allows all the logic gates of the deign at the same level of hierarchy. This allows the compiler to have better optimization and better area utilization for the design. If the hierarchical design is large, then this option may not work out. If number of hierarchies in the design increases, then compiler needs the larger amount of time during the design optimization phase.

Use the following command to achieve the logical flattening for the design

```
dc_shell> ungroup –all -flatten
dc_shell> compile –map_effort high –incremental mapping
dc_shell> report_timing –path full –delay max –max_path 1 –nworst 1
```

13.4.3 Use of group_path Command

The design performance can boost up to 10% by using the ***map_effort high*** option. But if timing is not met with the incremental compilation for the specified design constraints, then it is essential to group the ***critical timing paths*** and use the weight factor to boost the design performance. This command is useful to improve the timing performance. The command is shown below

```
dc_shell> group_path –name critical1 –from <input_name> –to
<output_name> –weight <weight factor>
```

Consider the design scenario which has the setup violation of 0.38 ns. The setup violation is the difference between the data required time and data arrival time. So, the slack is negative and setup time is violated.

```
dc_shell> read –format Verilog combinational_design.v
dc_shell> create_clock –name clk –period 15
dc_shell> set_input_delay 3 –clock clk in_a
dc_shell> set_input_delay 3 –clock clk in_b
dc_shell> set_input_delay 3 –clock clk c_in
dc_shell> set_output_delay 3 –clock c_out
dc_shell> current_design = combinational_design
dc_shell> compile –map_effort medium
```

```
dc_shell> report_timing –path full –delay max –max_path 1 –nworst 1
```

After the design is synthesized successfully, use the **report_timing** command. The timing report for the design is obtained with the multiple options as shown in the above script and shown in Example 1.

Example 1 Timing report with negative slack

```
Startpoint: c_in (input port)
Endpoint: c_out (output port)
Path Group: clk
Path Type: max
Point                                  Incr          Path
-------------------------------------------------------------
input external delay                   0.00          0.00 f
c_in (in)                              0.00          0.00 f
U19/Z (AN2)                            0.87          0.87 f
U18/Z (EO)                             1.13          2.00 f
add_8/U1_1/CO (FA1A)                   2.27          4.27 f
add_8/U1_2/CO (FA1A)                   1.17          5.45 f
add_8/U1_3/CO (FA1A)                   1.17          6.62 f
add_8/U1_4/CO (FA1A)                   1.17          7.80 f
add_8/U1_5/CO (FA1A)                   1.17          8.97 f
add_8/U1_6/CO (FA1A)                   1.17          10.14 f
add_8/U1_7/CO (FA1A)                   1.17          11.32 f
U2/Z (EO)                              1.06          12.38 f
C_out (out)                            0.00          12.38 f
data arrival time                                    12.38 f

clock clk (rising edge)                15.00         15.00
clock network delay (ideal)            0.00          15.00
output external delay                  -3.00         12.00
data required time                                   12.00
-------------------------------------------------------------
Data required time                                   12.00
Data arrival time                                    -12.38
Slack (violated)                               -0.38
```

To fix the setup violation and to boost the design performance, the designer can use the group_path with the weight factor. More the weight factor, more is the compilation time.

dc_shell> group_path –name critical1 –from c_in –to c_out –weight 8
dc_shell> compile –map_effort high –incremental mapping
dc_shell> report_timing –path full –delay max –max_path 1 –nworst 1

The above commands generate the timing report with positive slack and remove setup violation as shown in Example 2.

Example 2 Timing report with the positive slack

```
Startpoint: c_in (input port)
Endpoint: c_out (output port)
Path Group: max
Path Type: max
Point                           Incr        Path
----------------------------------------------------
input external delay            0.00        0.00 f
c_in (in)                       0.00        0.00 f
U19/Z (AN2)                     0.87        0.87 f
U18/Z (EO)                      1.13        2.00 r
add_8/U1_1/CO (FA1A)            2.27        4.27 f
add_8/U1_2/CO (FA1A)            1.17        5.45 f
add_8/U1_3/CO (FA1A)            1.17        6.62 r
add_8/U1_4/CO (FA1A)            1.17        7.80 f
add_8/U1_5/CO (FA1A)            1.19        8.99 r
add_8/U1_6/CO (FA1A)            1.15        10.14 f
add_8/U1_7/CO (FA1A)            0.79        10.93 f
U2/Z (EO)                       1.06        11.99 f
C_out (out)                     0.00        11.99 f
data arrival time                           11.99 f

clock clk (rising edge)         15.00       15.00
clock network delay (ideal)     0.00        15.00
output external delay           -3.00       12.00
data required time                          12.00
----------------------------------------------------
Data required time                          12.00
Data arrival time                           -11.99
----------------------------------------------------
Slack (met)                           0.01
```

As shown in the above timing report for the max analysis with the compile_map high option and weight factor of 5, the slack is met.

13.4.4 Submodule Characterizing

In the practical ASIC designs, the design can have multiple hierarchies. Consider that the top-level design consists of submodules X, Y, Z.

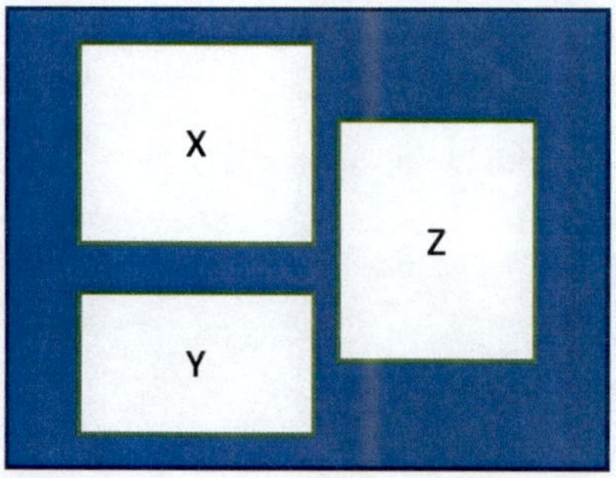

If the synthesis is performed for the individual blocks, then the timing can meet. When these submodules are instantiated in the top, then it may be possible that they do not meet the timing. That is, block-level timing met but the top-level timing fails.

The reason for this may be the glue logic used in between the submodules X, Y, Z or the improper partitioning at the top-level design hierarchy.

Under such circumstances to meet the design constraints, it is advisable to use the **characterize** command. This command allows the capturing of the boundary conditions for the submodule based on the top-level hierarchy environment. **Each submodule can be compiled and can characterize independently.**

The following is the script which can be used. Consider the submodule X, Y, Z instance names as I1, I2, and I3.

```
dc_shell> current_design = TOP
dc_shell> characterize I1
dc_shell> compile –map_effort high –incremental mapping
dc_shell> current_design = TOP
dc_shell> characterize I2
dc_shell> compile –map_effort high –incremental mapping
dc_shell> current_design = TOP
dc_shell> characterize I3
dc_shell> compile –map_effort high –incremental mapping
dc_shell> current_design = TOP
```

13.4.5 Register Balancing

Register balancing is efficient and powerful technique to have the pipelined stage. This technique improves the design performance by moving the logic and hence reduces the register-to-register delay. Consider the pipelined design shown in Fig. 13.10 and consisting of the three flip-flops and combinational logic.

Depending on the logic density of the combinational design, the arrival time may be different in the first and second reg-to-reg path. This can be balanced using the logic split technique by retaining the same functionality. To improve the design performance, it is recommended to use the additional pipelined stages.

The techniques like register balancing can be used to split the combinational logic from one of the pipelined stages to another pipelined stage without affecting the functionality of the design. This is achieved by compiler by using the following set of commands

```
dc_shell> balance_registers
dc_shell> report_timing -path full -delay max -max_path 1 -nworst 1
```

13.5 FSM Optimization

For the optimization of the finite state machines, the FSM compiler is used. The use of FSM compiler is with objective to optimize for the area and to improve the design performance. In the practical ASIC designs, the state machines are always coded as a separate module block. The FSM designs should have the clean data and timing path. It is recommended to use the suitable coding style while designing the state machine. The state machines should have the glitch-free output.

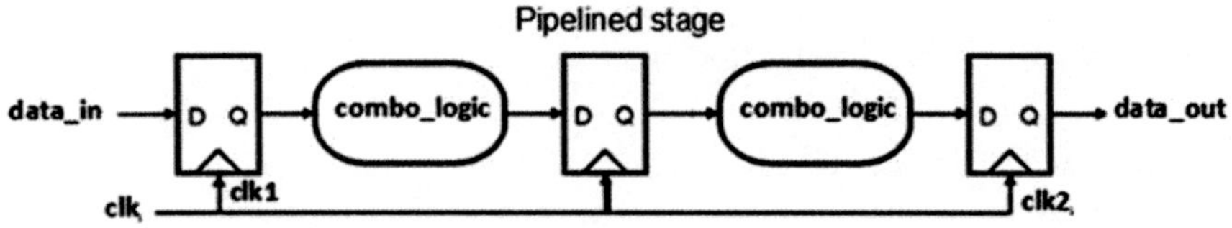

Fig. 13.10 Pipelined stages

The following script (Example 3) can be used for the FSM extraction and optimization.

Example 3 FSM extraction script

```
/* read the design object */
dc_shell> read -format verilog state_machines.v
/* Map the design */
dc_shell> compile –map_effort medium
/* if the design is not partitioned then group the logic */
dc_shell> set_fsm_state_vector { <flip_flop_name>, <flip_flop_name>,...}
dc_shell> group –fsm –design_name <fsm_design_name>
/* extract the state machine from netlist in the state machine table
format */
dc_shell> set_fsm_state_vector { <flip_flop_name>, <flip_flop_name>,...}
dc_shell> set_fsm_encoding { "state0=0", "state1=1", .......}
dc_shell>extract
/* write the design in the FSM format */
dc_shell>write –format  st –output state_machine.st
/* if the design is already in the state machine format then read the
design */
dc_shell>read –format  st  state_machine.st
/* define the order of the state */
dc_shell>set_fsm_order {state0,state1,....}
/* define the encoding style */
dc_shell> set_fsm_encoding_style <encoding_style>
/* compile the design */
dc_shell> compile –map_effort high
```

13.6 Fixing Hold Violations

To fix the setup violations, it is essential to modify the architecture of the design and in turn it has greater impact on the RTL coding of the design. The setup violations are fixed during the pre-layout STA, and hold violations can be fixed during post-layout STA phase as routing information is available after P and R. To fix the hold violations use the following command

```
dc_shell> set_fix_hold clk1
dc_shell> compile –map_effort_high- incremental_mapping
```

13.7 Report Command

The following are few commands used to generate reports.

13.7.1 report_qor

This is used to generate report which consists of timing summary of all the path groups. This gives overall status of the timing for the design. Example 4 shows the sample report with multiple timing path groups using ***report_qor*** command.

Example 4 qor report

```
Timing Path Group 'clk1'
------------------------------------------------------------------
Levels of Logic:                        6.00
Critical Path Length:                   3.64
Critical Path Slack:                    -2.64
Critical Path Clk Period:       11.32
Total Negative Slack:                   -55.45
No. of Violating Paths:                  59.00
No. of Hold Violations:                 1.00
------------------------------------------------------------------
Timing Path Group 'clk2'
------------------------------------------------------------------
Levels of Logic:                        10.00
Critical Path Length:                   3.59
Critical Path Slack:                    -0.29
Critical Path Clk Period:       22.65
Total Negative Slack:                   -2.90
No. of Violating Paths:                 11.00
No. of Hold Violations:                  0.00
------------------------------------------------------------------
Cell Count
------------------------------------------------------------------
Hierarchical Cell Count:                1736
Hierarchical Port Count:                          114870
Leaf Cell Count:                        323324
```

13.7.2 *report_constraints*

This command is used to report the constraints. The following is the report generated using the **report_*constraints*** command.

Example 5 Report constraints

Weighted Group (max_delay/setup)	Cost	Weight	Cost
CLK	0.00	1.00	0.00
default	0.00	1.00	0.00
max_delay/setup	0.00		
Constraint	Cost		
max_transition	0.00 (MET)		
max_fanout	0.00 (MET)		
max_delay/setup	0.00 (MET)		
critical_range	0.00 (MET)		
min_delay/hold	0.40 (VIOLATED)		
max_leakage_power	6.00 (VIOLATED)		
max_dynamic_power	14.03 (VIOLATED)		
max_area	48.00 (VIOLATED)		

13.7.3 *report_contraints_all*

This command is used to show all the timing and DRC violations. The report (Example 6) is generated using the ***report_constraints_all*** command.

Example 6 All constraint report

```
max_delay/setup ('clk1' group)
                 Required        Actual
Endpoint         Path Delay      Path Delay     Slack
----------------------------------------------------------------------------------
data[15]         1.00            3.64 f         -2.64 (VIOLATED)
data[13]         1.00            3.64 f         -2.64 (VIOLATED)
data[11]         1.00            3.63 f         -2.63 (VIOLATED)
data[12]         1.00            3.63 f         -2.63 (VIOLATED)
----------------------------------------------------------------------------------
```

Example 7 is the sample script and can be used to constrain the design for maximum operating frequency of 500 MHz.

Example 7 Sample script for constraining design at 500 MHz

```
/* set the clock */
set clock clk
/* set clock period */
set clock_period 2
/* set the latency */
set latency 0.05
/* set clock skew */
set early_clock_skew [expr $ clock_period/10.0]
set late_clock_skew [expr $ clock_period/20.0]

/* set clock transition */
set clock_transition [expr $ clock_period/100.0]
/* set the external delay */
Set external_delay [expr $ clock_period*0.4]
/* define the clock uncertainty*/
set_clock_uncertainty –setup $ early_clock_skew
set_clock_uncertainty –hold$ late_clock_skew

Name the above script as clock.src, and Source the above script

/* report clock and timing*/
dc_shell> report_timing
dc_shell> report_clock
dc_shell> report_timing
dc_shell> report_constraints –all_violations
```

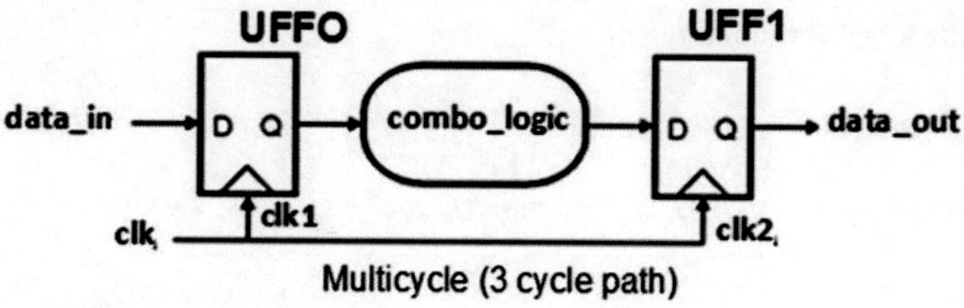

Fig. 13.11 Multicycle path

Table 13.4 Multicycle path commands

Command	Description
create_clock –name clk_master -period 5 [get_ports clk_master]	To create the master clock of period 5 ns
set_multicycle_path 3 –setup –from [get_pins UFFO/Q] –to [get_pins UFF1/D]	This sets the multicycle path of 3 cycles
set_multicycle_path 2 –hold –from [get_pins UFFO/Q] –to [get_pins UFF1/D]	This is used to move the hold check to 2nd clock cycles and setup is checked at 3rd clock cycles

13.8 Multicycle Paths

The multicycle paths in the design need to be reported as they are timing exceptions. The paths can be set so that the timing analyzer can perform the setup and hold check (Fig. 13.11).

The commands used to set the multicycle path are listed in Table 13.4.

Consider the design which has the complex multipliers, and they need to have few clock cycle times as their inputs and outputs are registered. Here assumption is that, we have registered inputs and outputs! In such circumstances, we need to set the multicycle path (Fig. 13.12).

13.9 Chapter Summary

The following are important points to conclude the chapter

1. Avoid use of the combinational logic as individual block or module.
2. Do not use the glue logic between two modules.
3. Use set_max_area attribute while synthesizing the design.
4. For the optimization of the finite state machines, the FSM compiler is used.
5. The design hierarchy of the design can be broken by using logical flattening of the design.

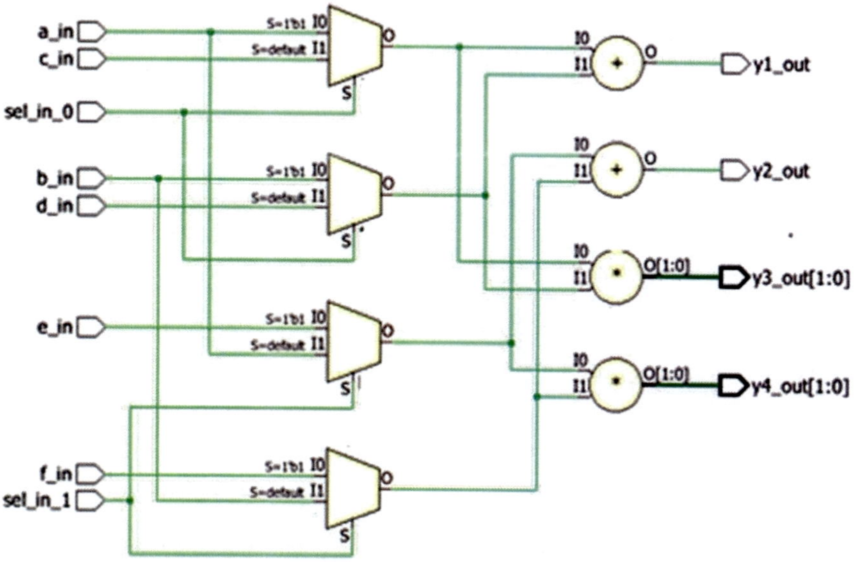

Fig. 13.12 Complex multipliers as combinational logic in reg-to-reg path

6. If timing is not met with the incremental compilation by using the design constraints, then it is essential to group the critical timing paths and use the weight factor to boost the design performance.

Chapter 14
Design for Testability

The logic design and synthesis phase give the lower level of abstraction for the design which is gate-level netlist. To detect the early faults in the design, the DFT team needs to work on the strategies during the architecture and at logic levels. The strategies may be having the DFT-friendly RTL and DFT-friendly architecture design. Deepening on the time and budget requirement for any kind of complex ASIC design, the faults like stuck at '1', '0', and memory-related faults can be detected.

The chapter discusses about the various faults and strategies used during the DFT.

14.1 What Is Need of DFT?

Let us try to address the question that why we need go for the DFT? Following are few of the important points to address the actual issues after chip is manufactured.

1. After chip manufacturing in the lot size of few millions, it is not guaranteed that all the chips will be functionally operated. That is there are few % of chips which will have the defects.
2. After manufacturing, we will be able to see only the inputs and outputs of the chip and internal node and pin information is not available, and in such scenarios if the faults are within the chip, then it is very huge loss to the design houses.

Now, let us try to understand that why faults will remain in the design? Consider that the design functionality is correct and even the verification results indicate the functional correctness for the design. Still during routing due to congestions and due to power routing if the output of cell is shorted to Vdd or Vss, then the cell will have permanently pull-up or pull-down state which we call as stuck at fault.

DFT and scan insertions are used to detect early faults in the designs.

V. Taraate, *ASIC Design and Synthesis*,
https://doi.org/10.1007/978-981-33-4642-0_14

So, detect these faults during early stage of the design, the DFT techniques and strategies are useful. We need to have the test patterns to test the chip and that's what we will try to do during the DFT.

14.2 Testing for Faults in the Design

The real scenario is the physical testing is carried out after chip is manufactured. But if the faults remain in the design, the overall lot is rejected. We carry out the verification for the design and testing at the device level. This fault can be perceived during the early stage of designs, and for that, we use the DFT.

The SFT increases the area and cost for the design, but we have the fault coverage for the chip. Defects can be physical and electrical.

Physical defects are due to silicon defects may be due to defective oxide.

Electrical defects can be short, open, transition or change in the threshold voltage.

Faults in the design: Following can be the types of faults in the design

1. **Crosspoint faults**: It may be due to the deficiency or due to extra metal.
2. **Bridging faults:** Input bridging or output bridging faults.
3. **Transition Faults:** Output is not changing with input change.
4. **Delay faults:** Due to the gate slow paths.
5. **Stuck at fault**: The output is stuck at '0' (short) or stuck at '1' (pulled up) permanently.
6. **Pattern-sensitive faults in the memories**: Faults in the memories and during DFT we need to have the various memory fault detection techniques.

14.3 Testing

Effectively we should have the test protocol during test, and it should have

1. Test pattern generation
2. Application of the test patterns
3. Evaluation.

So, the test flow is basically using the four important steps

1. Identification of the faults
2. Test generation
3. Fault simulation
4. Design testability
5. DFT.

14.4 Strategies Used During the DFT

Consider that we have performed the logic synthesis, what exactly we have done is that we got the gate-level netlist using non-scan cells if the RTL and design architecture are not DFT friendly. So, the real demand and strategy during ASIC design is use of the

1. DFT-friendly architecture
2. DFT-friendly RTL.

As during the top- or block-level verification, the faults are not detected, and better way is to perform the test synthesis using scan methodology to improve the controllability and observability for the design. Thus, in simple words, we can say that the scan cells with suitable scan method are useful to detect the faults during the DFT.

Effectively we will try to replace all or few non-scan sequential cells using the scan cells, where the scan cells are connected to form the shift register which is shown in Fig. 14.1.

The important terms which we need to understand are the controllability and observability for the node.

As shown in Fig. 14.2, the D input of flip-flop is not controllable and observable and needs to have the provisions so that the testing can be performed. Each node should be controllable and observable.

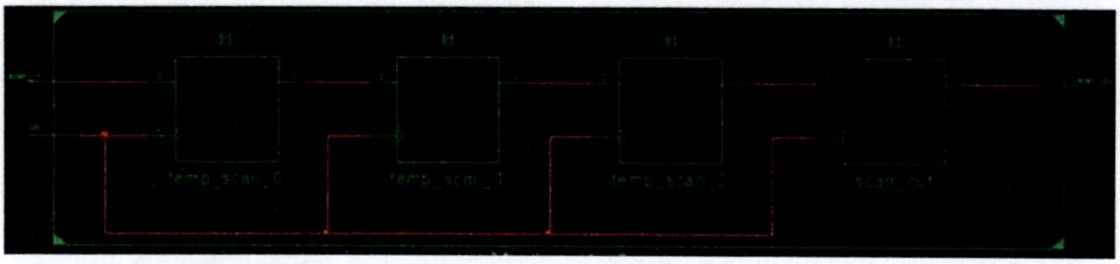

Fig. 14.1 Scan chain

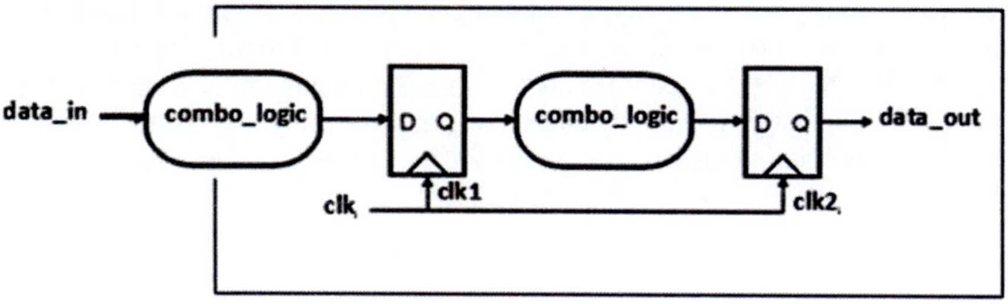

Fig. 14.2 Design without controllability and observability

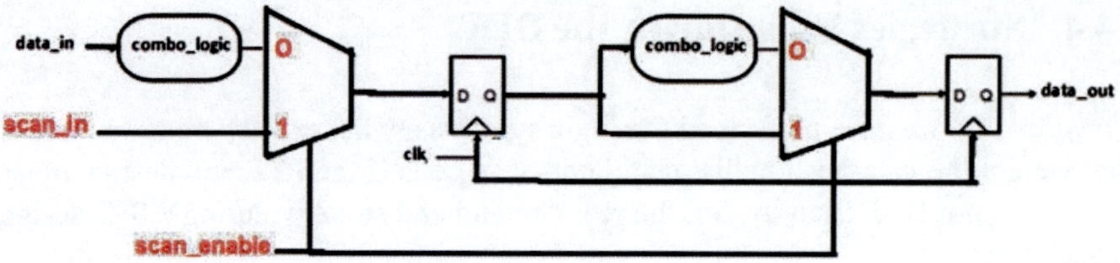

Fig. 14.3 Design with each node controllable and observable

Controllability: It is the ability to control the nodes of the sequential circuit by a set of the inputs. Each node in the sequential circuit should be controllable that means the test vector should be able to reach to that node during scan mode.

Observability: How efficiently we can observe the change at the nodes indicates the observability. That is, we should get the desire change of the state at the outputs.

So, during the scan, each node should be controllable and observable. For scan_enable = 1, the scan_in passes through the D input of each multiplexer, and it indicates that the design has mux-based scan chain(Fig. 14.3) where each node is controllable and observable. For more details, refer next few sessions.

14.5 Scan Methods

Depending on the DFT strategies, time and budgeting, the scan methods are chosen. The scan methods we can use as

1. ***Full Scan***: In this all the sequential cells replaced by scan cells. This has higher fault coverage.
2. ***Partial Scan***: In this, few of the sequential cells are replaced by scan cell. Fault coverage using this method is function of the number of sequential cells replaced by scan cells.

Having strategy to go for the full scan or partial scan is based on the area and timing constraints. Consider that we have the floating-point engine which has remarkably high density. For such kind of designs, after placement and routing the congestion may be more and as the design demands more area the possibility of the stuck at fault may be on higher side. So, by considering this, we can go for the full scan for the floating-point unit. Full scan is used to have the higher fault coverage.

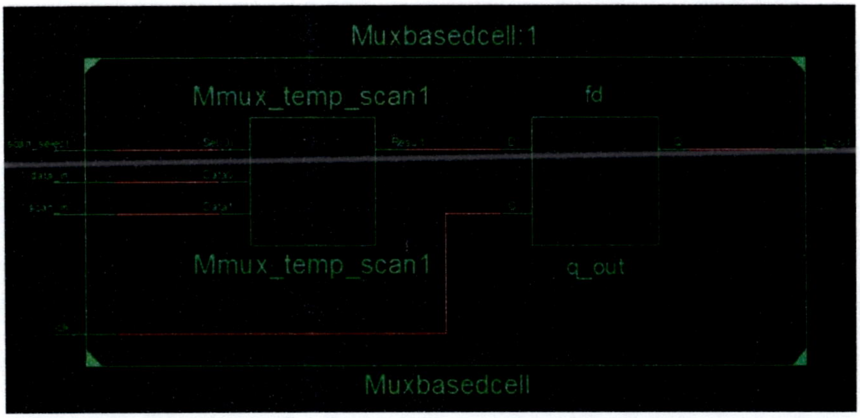

Fig. 14.4 Mux-based scan cell

14.5.1 Mux-Based Scan

As shown in Fig. 14.4, the mux-based scan methodology uses the multiplexer, D flip-flop and due to simple mechanism it is popular in the industry. Due to use of the multiplexer at the flip-flop selection input, the design operates in the functional mode (normal mode) or in the scan mode (test mode). That is the flip-flop input is controllable.

14.5.2 Boundary Scan

JTAG is popular as the protocol for the boundary scan tests. We have most of the time JTAG-based test schemes for the FPGAs and ASICs (Fig. 14.5).

14.5.3 Built-In Self-Test (BIST)

The memory BIST (MBIST) and Logic BIST(LBIST) are used to have the built-in test features in the ASIC.

The MBIST is shown in Fig. 14.6 and useful to incorporate the test features for the chip.

In this book, we will try to understand the mux-based scan chain.

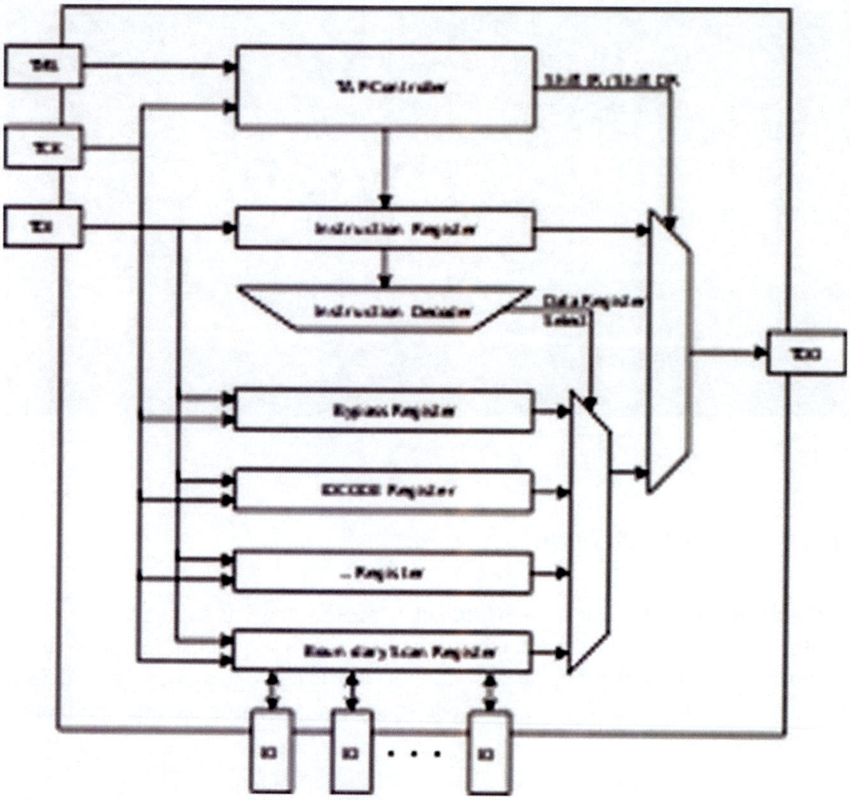

Fig. 14.5 JTAG IEEE 1149.1 standard for boundary scan

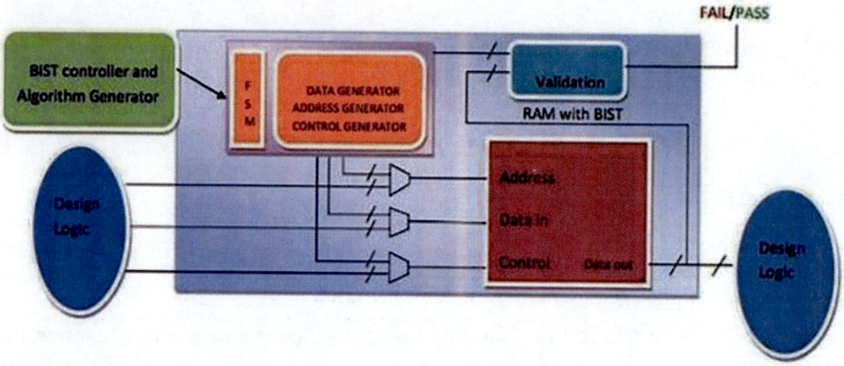

Fig. 14.6 Memory BIST

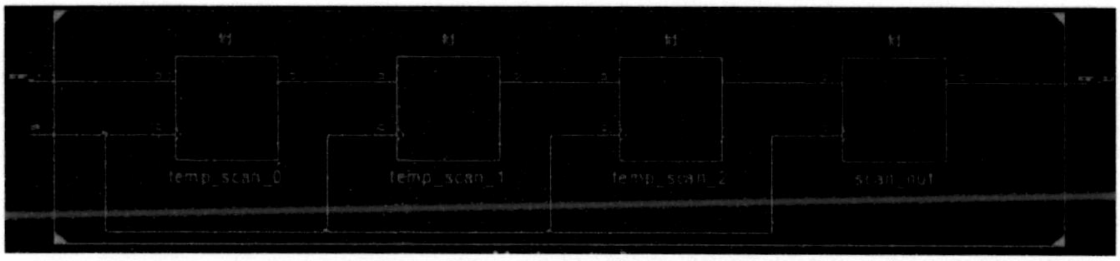

Fig. 14.7 Shift register as scan chain

14.6 Scan Insertion

What is the real advantage of the scan chain is that it is easy to pass the test pattern using the scan chains? If we have 16 inputs, then number of test patterns can be 2 to the power 16 which is extremely high and time consuming without the use of the scan chain. Scan insertion means use of the scan cells (shift register) as shown in Fig. 14.7 so that all the design nodes are controllable and observable.

The advantage of the scan chain is that it reduces the overall time during the testing. During normal mode, the scan cell should behave same as the normal sequential cell. During the scan, the scan data is shifted serially, and depending on the number of scan cells used in the chain, it takes those many number of cycles. In Fig. 14.7, each scan element is mux-based scan cell.

This technique increases the area and introduces the delay in the design. Now, as scan insertion is performed, how we can test the design? Let us generate the test pattern and use to test the design.

14.7 Challenges During the DFT

Following are important challenges during the DFT

1. How many number of clocks the design has and how many test clock design support?
2. What kind of the tester is used and whether it supports waveforms?
3. How many scan chains we need to insert and what is length of each scan chain?
4. If the design has area limitations, then the sharing of the scan ports with the functional ports is possible or not?
5. What is test vectors size and how many numbers of scan bits for the design?

These challenges need to be addressed. As stated earlier that due to scan insertions, it significantly affects on the area of the design.

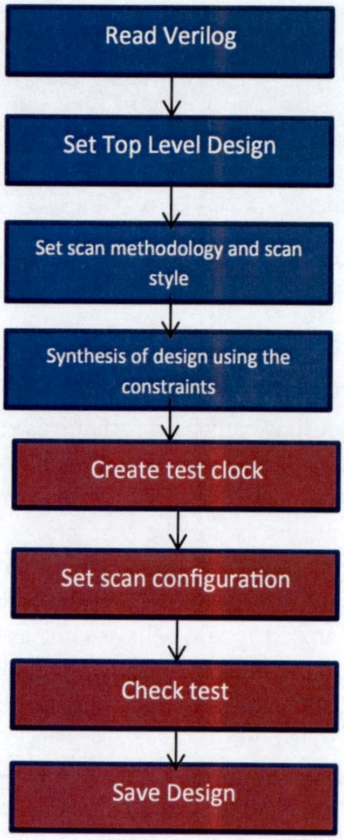

Fig. 14.8 DFT flow steps-1

14.8 DFT Flow and Test Compiler Commands

Use the DFT flow shown in Figs. 14.8 and 14.9 to generate test vectors

Table 14.1 describes few of the test compiler commands and their use during DFT

14.9 The Scan Design Rules to Avoid DRC Violations

Few of the design guidelines and scan design rules to avoid DRC violations are listed in this section.

1. There should not be any combinational loop in design (Fig. 14.10).
   Solution: For this design, solution to avoid DRC is break the combinational loop using the control element.

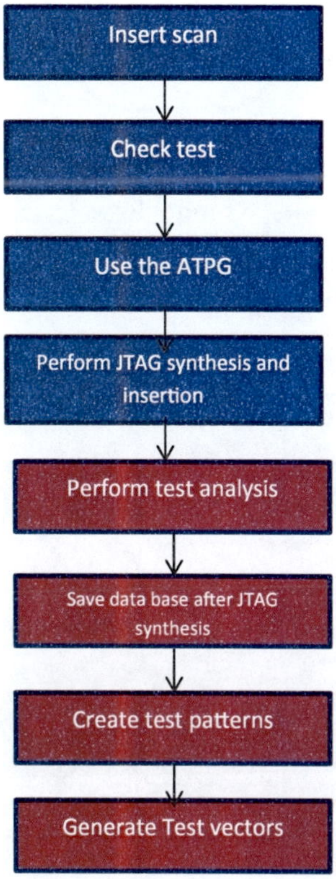

Fig. 14.9 DFT flow steps-2

Table 14.1 Important test compiler commands

Command	Description
check_test	It is used to infer the test protocol to perform the DRC check k by simulating the test protocol
create_test_clok	It is like create_clock command which we have used using DC and used to specify the waveform and period of the test clock
insert_scan	To form the scan chain by replacing the sequential cells using the scan cells
create_test_pattern	To create the test patterns and to have the binary vectors, the command is used
set_test_hold	To specify the static values in the primary ports
set_test_dont_fault	Used to remove the specific faults. For example, the memory tests are different as compared to stuck at fault test

2. Avoid use of latches (Fig. 14.11)
   Solution: try to have strategy to enable latches during the scan mode.
3. Try to avoid the generated clocks

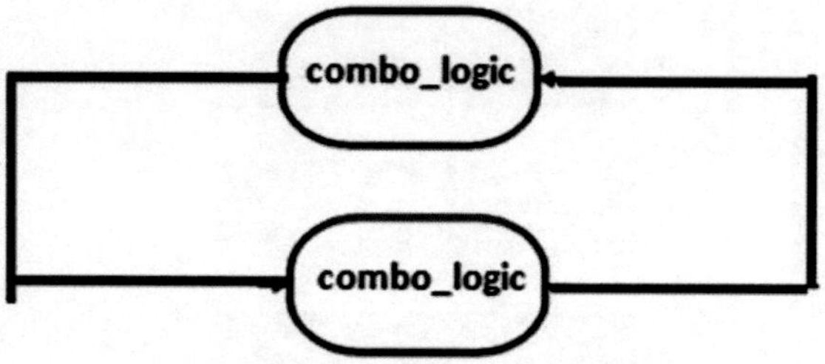

Fig. 14.10 Combinational loop in design

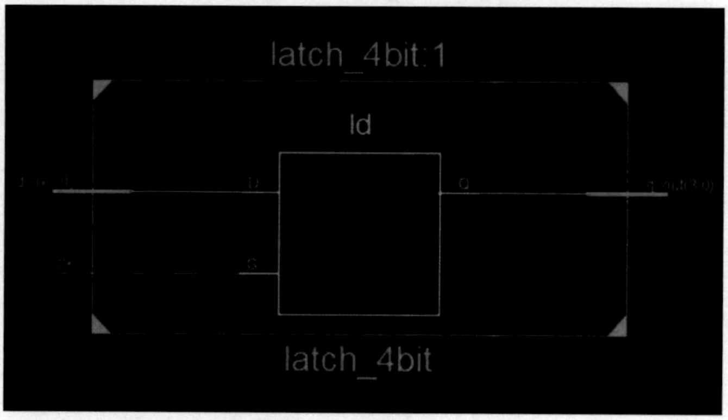

Fig. 14.11 Latch-based designs

4. Use the port-level clock signal to control the internally generated clocks.
5. Avoid the use of internally generated reset signals (Fig. 14.12).
   Solution: Use the reset control from main port to control the asynchronous reset signals.
6. There should not be gated clocks in the design (Fig. 14.13)
   Solution: Enable gated clocks in the scan mode.
7. Have a strategy for not replacing the shift registers during scan.
8. Use the techniques to bypass memories.
9. Avoid the use of positive and negative edge triggered flip-flops in single module
   Solution: using the mux-based glue logic invert the clock of negative edge triggered element during scan mode
10. Avoid mixed edge triggering (Fig. 14.14).

There are various other techniques useful to detect the faults, and for more detail, readers can refer the DFT books and test compiler manuals.

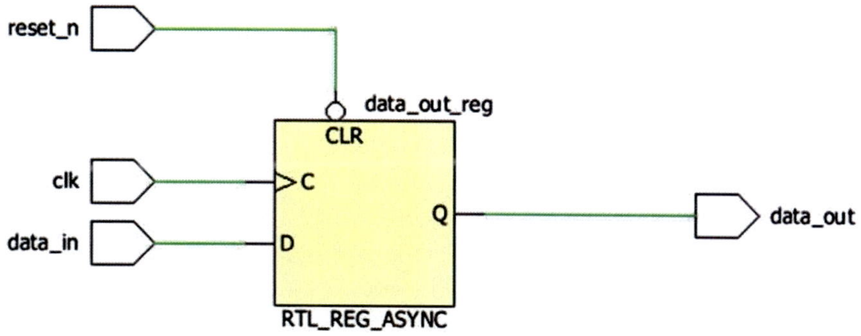

Fig. 14.12 Internally generated reset

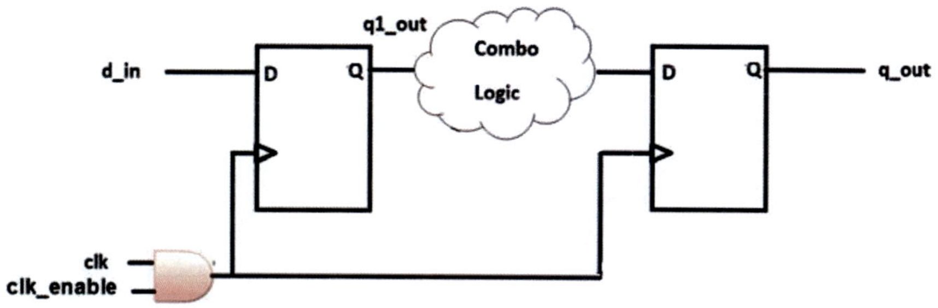

Fig. 14.13 Gated clocks in the design

Fig. 14.14 Mix of clock edges

14.10 Chapter Summary

Following are important points to conclude the chapter.

1. DFT is useful to detect early faults in the design.
2. Full scan indicates all the flip-flops in the design are replaced by scan cells.
3. Partial scan indicates that few of the flip-flops are replaced by scan cells.
4. Have the DFT-friendly architecture and RTL.
5. Stuck at faults are the due to net stuck at 0 or stuck at 1.

Chapter 15
Timing Analysis

The speed of the ASIC design is very important criteria, and the constraints for the block-level and top-level design are specified during the synthesis. These constraints should meet under any circumstances which indicate the clean design timing. The timing analysis for the ASIC designs is performed during the logical design with the following goals

1. To understand where design fails during logic level as the routing information is not available.
2. Is the design has the timing exceptions?
3. How many timing paths in the design violates and what should be strategies to fix the setup violations.

During the post-layout STA as the clock tree, network, routing delay information is available, the STA is performed to fix the timing violations that is setup and hold time violations. Various strategies are useful to fix the timing violations, and few of them are discussed in this chapter.

15.1 Introduction

Timing analysis tool uses the design constraint file and the vendor libraries with the timing information to perform the timing analysis for the design. Timing analysis is of two types static and dynamic.

Static Timing Analysis (STA): It is performed without use of any set of vectors with the goal to report the timing paths having the timing violations that are setup and hold violations. It is vector less approach.

The timing analysis for the ASIC design need to be performed during pre-layout and post-layout stage.

V. Taraate, *ASIC Design and Synthesis*,
https://doi.org/10.1007/978-981-33-4642-0_15

Dynamic Timing Analysis (DTA): The DTA is performed by use of the set of vectors for the design. The goal is to fix the set-up and hold time violations for the design.

Role of EDA Tool: The ***Synopsys PT*** is powerful timing analysis tool and used to report the timing paths which have positive and negative slack. If the setup and hold time for the design is not violated, then the design doesn't have the timing violation. Reporting of the timing paths which has timing violation is the real objective, and we can get the information from the timing report. The timing analysis tool or timing analyzer is used to report the overall timing information for the design. The team should have the strategies to fix these timing violations during the pre-layout and post-layout stage.

15.2 What Are Timing Paths for Design

As discussed in Chap. 6, for any kind of synchronous sequential design, we can have one or more than one timing path. Consider the design shown in Fig. 15.1.

The design has mainly four timing paths, and they are named as

1. Input to reg path
2. Reg to output path
3. Reg to reg path
4. Input to output path.

To identify the timing paths in the design, the designer should know the start point and end point.

Start Point: The clock input port of the sequential element (clk), data inputs (primary ports) of the sequential design is treated as the start point, and the tool algorithm identifies initially the start points for the design and then end points.

End Point: The end point of the design is the output port of the sequential element or data input of the sequential element D flip-flop (D).

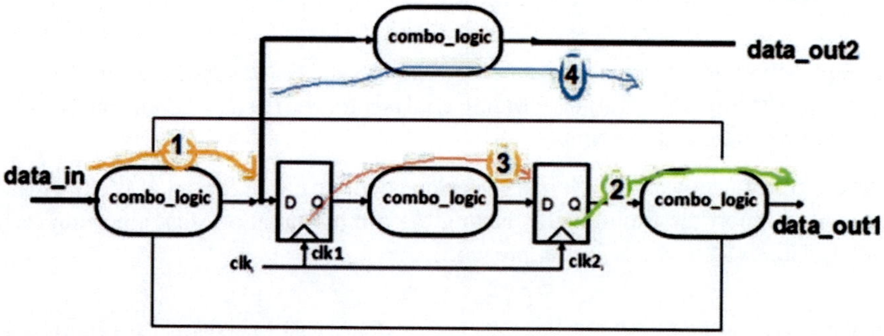

Fig. 15.1 Synchronous design

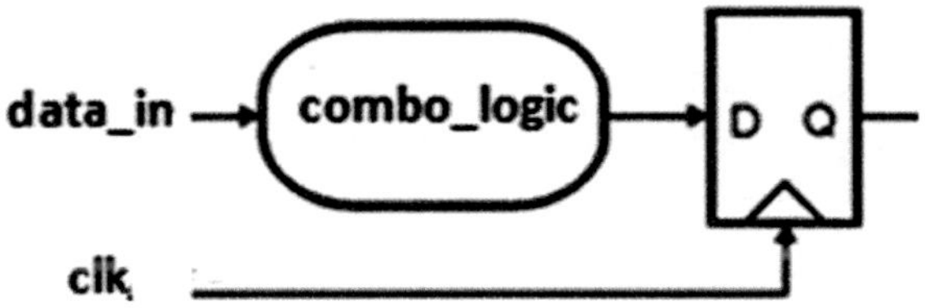

Fig. 15.2 Input to reg path

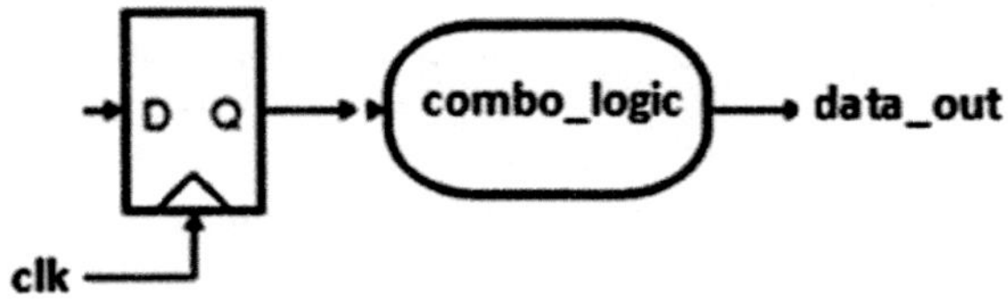

Fig. 15.3 Reg to output path

15.2.1 Input to Reg Path

It marked as path 1 in the figure, and it is from the input port data_in of the design to the D input of the sequential element (Fig. 15.2).

15.2.2 Reg to Output Path

It marked as path 2 in the figure, and it is from the clock pin clk2 of the flip-flop to the data_out1 of the sequential element (Fig. 15.3).

15.2.3 Reg to Reg Path

It marked as path 3 in the figure, and it is from the clock pin clk1 of the flip-flop to the data input of the D flip-flop that is sequential element 2 (Fig. 15.4).

Fig. 15.4 The reg to reg path

15.2.4 *Input to Output Path*

It is unconstrained path and also called as combinational path. It is marked as path 4, and it is from data_in of the design to the data_out2 of the design (Fig. 15.5).

15.3 Let Us Specify the Timing Goals

Consider the design shown in Fig. 15.6, and we need to perform the timing analysis for the single clock domain designs then what we need to do? Effectively, we need to specify the timing goals that are information about the clock and what is the maximum operating frequency for the design? What are IO delays and the skew information to the timing analysis tool?

Block-level STA: Let us use the following steps and try to create the script which can be used to perform the timing analysis.

1. Specify the clock
2. Specify the clock latency
3. Specify the setup uncertainty
4. Specify the hold uncertainty
5. Specify the input delay
6. Specify the output delay.

Fig. 15.5 Combinational path

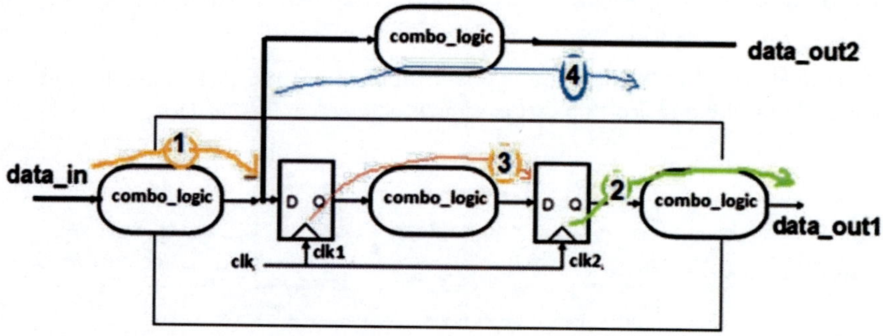

Fig. 15.6 Synchronous sequential design

- *Clock definition*:

For the processor top module, let us define the clock of 500 MHz having 50% duty cycle using SDC command

```
create_clock -period 2.00 -name clk [ get_ports {clk} ]
```

The above SDC command generates the clock of 500 MHz with the 50% duty cycle

- ***Specify clock latency***

Let us specify the clock latency. For example, if the clock latency is of 0.25 ns, then specify using the command 'set_clock_latency'

```
set_clock_latency -source 0.25 [ get_clocks clk]
```

- ***Specify early clock catency***

The early clock latency can be specified by the following SDC

```
set_clock_latency -source -early -rise -0.10 [ get_clocks clk]

set_clock_latency -source -early -fall -0.05 [ get_clocks clk]
```

- ***Specify the setup uncertainty***

Specify the uncertainty for the setup using command

```
set_clock_uncertainty -setup 0.5 [ get_clocks clk]
```

- ***Specify the hold uncertainty***

Specify the uncertainty for the hold using command

```
set_clock_uncertainty -hold 0.25 [ get_clocks clk]
```

- ***Specify the minimum input delay***

Specify the minimum input delay for the design using 'set_input_delay' command

```
set_input_delay -clock clk -min 0.1 find (port
data_in)
```

- ***Specify the maximum input delay***

Specify the minimum input delay for the design using 'set_input_delay' command

```
set_input_delay -clock clk -max 0.15 find (port
data_in)
```

- ***Specify the minimum output delay***

Specify the minimum output delay for the design using 'set_output_delay' command

```
set_output_delay -clock clk -min 0.1 find (port
data_out1)
```

- ***Specify the maximum output delay***

Specify the maximum output delay for the design using 'set_output_delay' command

```
set_output_delay -clock clk -max 0.15 find (port
data_out1)
```

15.4 Timing Reports

Perform the timing analysis for the design and then use the command **report_timing** to get the timing report for the design. The sample timing report is shown in Example 1 and for the design the slack is not met that is the design has timing violations as setup slack is negative.

Example 1 Sample timing report with negative slack

```
Startpoint: c_in (input port)
Endpoint: c_out (output port)
Path Group: clk
Path Type: max
Point                                   Incr                Path
--------------------------------------------------------------------------
input external delay                    0.00                0.00 f
c_in (in)                       0.00                0.00 f
U19/Z (AN2)                             0.87                0.87 f
U18/Z (EO)                              1.13                2.00 f
add_8/U1_1/CO (FA1A)            2.27                4.27 f
add_8/U1_2/CO (FA1A)            1.17                5.45 f
add_8/U1_3/CO (FA1A)            1.17                6.62 f
add_8/U1_4/CO (FA1A)            1.17                7.80 f
add_8/U1_5/CO (FA1A)            1.17                8.97 f
add_8/U1_6/CO (FA1A)            1.17                10.14 f
add_8/U1_7/CO (FA1A)            1.17                11.32 f
U2/Z (EO)                               1.06                12.38 f
C_out (out)                             0.00                12.38 f
data arrival time                                   12.38 f

clock clk (rising edge)             15.00           15.00
clock network delay (ideal)             0.00                15.00
output external delay                   -3.00               12.00
data required time                                          12.00
--------------------------------------------------------------------------
Data required time                                          12.00
Data arrival time                                           -12.38
Slack (violated)                                            -0.38
```

Using the strategies and performance improvement techniques try to fix the setup violation from the design. These strategies are discussed in the next subsequent section.

The timing report after fixing for the timing violations is shown in Example 2.

Example 2 Sample timing report with the positive slack

Startpoint: c_in (input port)
Endpoint: c_out (output port)
Path Group: max
Path Type: max

Point		Incr		Path
input external delay		0.00		0.00 f
c_in (in)	0.00		0.00 f	
U19/Z (AN2)		0.87		0.87 f
U18/Z (EO)		1.13		2.00 r
add_8/U1_1/CO (FA1A)	2.27		4.27 f	
add_8/U1_2/CO (FA1A)	1.17		5.45 f	
add_8/U1_3/CO (FA1A)	1.17		6.62 r	
add_8/U1_4/CO (FA1A)	1.17		7.80 f	
add_8/U1_5/CO (FA1A)	1.19		8.99 r	
add_8/U1_6/CO (FA1A)	1.15		10.14 f	
add_8/U1_7/CO (FA1A)	0.79		10.93 f	
U2/Z (EO)		1.06		11.99 f
C_out (out)		0.00		11.99 f
data arrival time			11.99 f	
clock clk (rising edge)	15.00		15.00	
clock network delay (ideal)		0.00		15.00
output external delay		-3.00		12.00
data required time				12.00
Data required time				12.00
Data arrival time				-11.99
Slack (met)			0.01	

As shown in the timing report, the slack is positive and the design doesn't have the setup violation.

15.5 Strategies to Fix Timing Violations

If the design has the setup and hold violations, then it is nightmare for the STA team to fix these violations. For any kind of ASIC design during the STA, we need to report all the paths having violations, and we should be able to have the strategies in place to fix these timing violations. Consider the design of the processor and we have to perform the block-level and top-level STA! (Fig. 15.7).

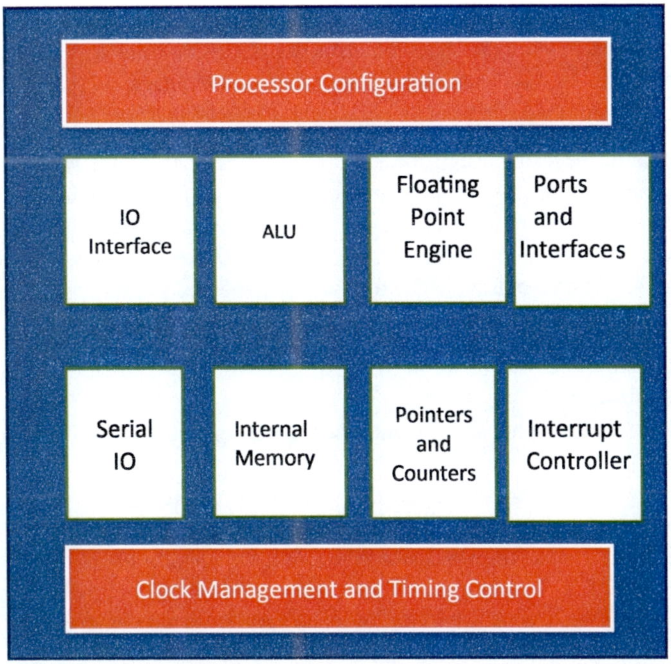

Fig. 15.7 Processor top-level architecture

The following are few of the issues and violations we can experience during the STA

1. The block-level timing met but the top-level timing has issue?
2. There are lot of timing paths in the floating point engine which has setup violations.
3. There are few timing paths for the general-purpose processing engines where we have data very fast have hold violations.
4. Due to large density multipliers the timing exceptions in the design.
5. At the asynchronous sequential boundaries the timing violations due to metastable output from synchronizer.

The following are few of the techniques useful to fix the timing violations

1. **Use tool-based directives**: Few of the tool-based directives to balance the delays for the register to register path can be useful during optimization phase. The register balancing can be useful to balance the delays by introducing the latency in the design.
2. **Resynthesize the design with goal of performance improvement**: Use the pipelining architecture, register balancing, and duplication techniques to fix the timing violations. If not able to fix for all the timing violations, then use the architecture and RTL tweaks.

3. **Architecture and micro-architecture Tweaks**: To fix the timing violations, the final option is to work on the architecture tweaks by incorporating the parallelism or pipelining. But, this has additional impact as the design cycle elongates.

Most of the time, during the STA, we may experience the scenario that the block-level constraints meet and design doesn't have timing violation. But while performing the timing analysis for the top-level design, the design has many timing violation. This may be due to the top-level integration and the improper design partitioning. In such scenario, it affects on the various timing paths as the delay introduced has significant impact on the design. The following can be good strategy to fix the timing violations in such circumstances

1. Try to have the closure look at the design partitioning and were the additional delays introduces.
2. Try to set the false path at the asynchronous multiple clock boundaries to disable timing.
3. Try to find the late arrival signals to fix the setup violations.
4. Try to find the early arrival signal and have the strategies to fix the hold violations.
5. Try to check for the timing exceptions and try to specify the timing exceptions in the Tcl script.

15.5.1 Fixing Setup Violations in the Design

The setup violations are due to the data or the control signal changes during the setup window. The reason being the logic density is higher, and it adds the significant amount of delay and data changes during the setup window prior to arrival of active edge of the clock. The following are few techniques used to fix the setup time violations:

1. Split the combinational delays by retaining design functionality
2. Use the encoding methods to avoid the cascade logic
3. Have a strategy to fix for the late arrival signals
4. Use register balancing or pipelining.

15.5.1.1 Duplicate the Logic

Consider we have the reg to reg path and have the timing setup violation. Then, the reason is the large amount of delays which has incurred due to combinational logic. In such scenario, use the logic duplication concepts to duplicate the logic. This has impact on the area, but due to parallelism, we will be able to fix for the setup violations (Fig. 15.8).

Fig. 15.8 Combo logic with large delay

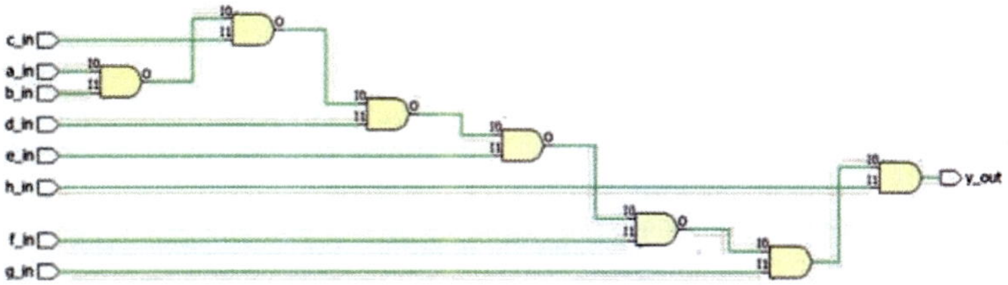

Fig. 15.9 Cascade or priority logic

15.5.1.2 The Priority Versus Multiplex Encoding Methods

The popular encoding methods are priority encoding and multiplexed encoding. Consider the continuous assignment to have the glue logic (Fig. 15.9).

```
assign y_out=a_in && b_in && c_in && d_in && e_in && f_in && g_in
&& h_in;
```

The ***assign*** construct infers the priority logic where the a_in has highest priority over the other inputs. Due to cascade AND gates, the glue logic has more delay and slower, and hence, there are chances of setup violations. By using the parallel logic that is by structuring and grouping the terms, the delay can be reduced. If each AND gate has 0.25 ns delay, then cascade stage has delay of 1.75 ns.

As shown in Fig. 15.10, if we use the parallel logic or mux kind of logic which has many parallel inputs, then the overall glue logic delay is just three stage delay which is 0.75 ns.

```
Use the assign

y_out= ( (a_in && b_in) && (c_in && d_in)) && ( (e_in &&
f_in) && ( g_in && h_in);
```

15.5.1.3 Late Arrival Signals

The setup time is violated due to late arrival signals and the design fails. Consider the design shown in Fig. 15.11.

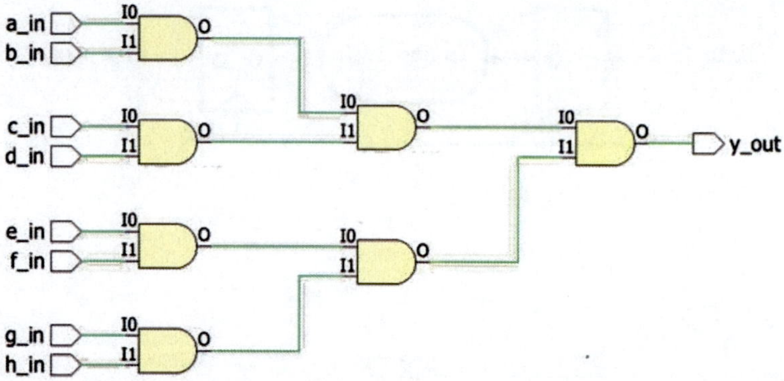

Fig. 15.10 Multiplexed or parallel logic

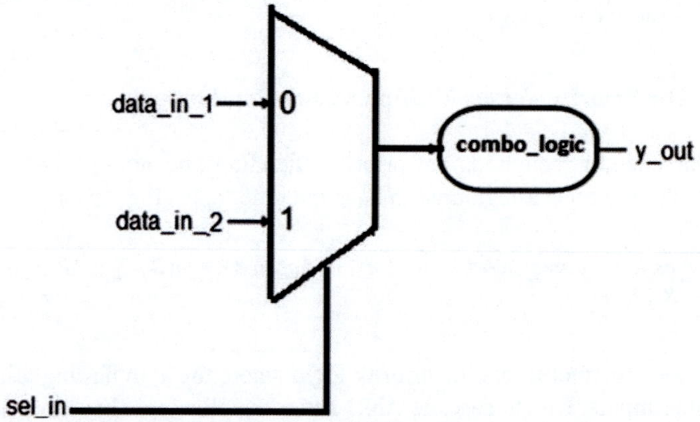

Fig. 15.11 Logic which has late arrival signal sel_in

As shown in the design, the glue logic is used to select from one of the inputs data_in_1 or data_in_2 and both arrives at the same time. But the sel_in input arrives late. As y_out is used as input at D flip-flop, the design has setup violation. That is data changes during the setup window.

The setup violation can be fixed using some strategy of moving the design blocks at the input side and using the logic duplication. As sel_in is late arrival time and it is the reason for the setup violation, we will duplicate the combinational logic (combo_logic) at the inputs of multiplexer so that margin of the combinational logic delay can be used to sample the late arrival signal and in turn to fix the setup violation. Figure 15.12 describes the strategy used using the RTL design tweak.

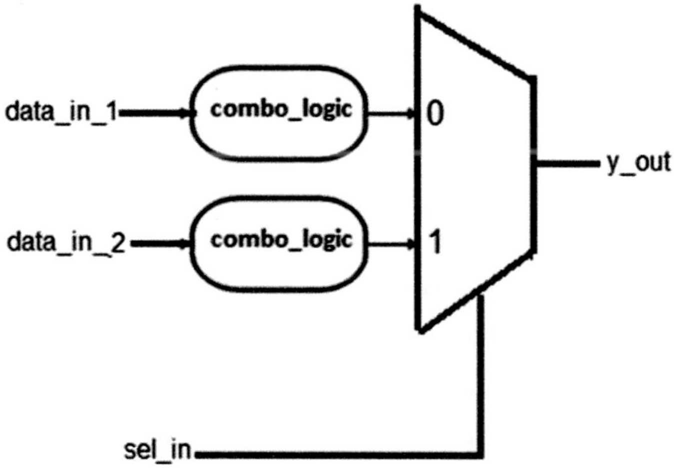

Fig. 15.12 Logic duplication which has late arrival signal sel_in

15.5.1.4 Register Balancing

To fix the set up time and to improve the design performance, register balancing is one of the powerful techniques. Consider the operating frequency for the design as 500 MHz; that is, clock time period is 2 ns. If reg to reg path has more combinational delay, then the data arrival is slow and the flip-flop goes into the metastable state and hence has the setup violation (Fig. 15.13).

So, use the pipelining or register balancing by splitting the combinational logic as shown in Fig. 15.14. Care should be taken that the strategy and tweaks should not effect on the design functionality.

Fig. 15.13 Logic without register balancing

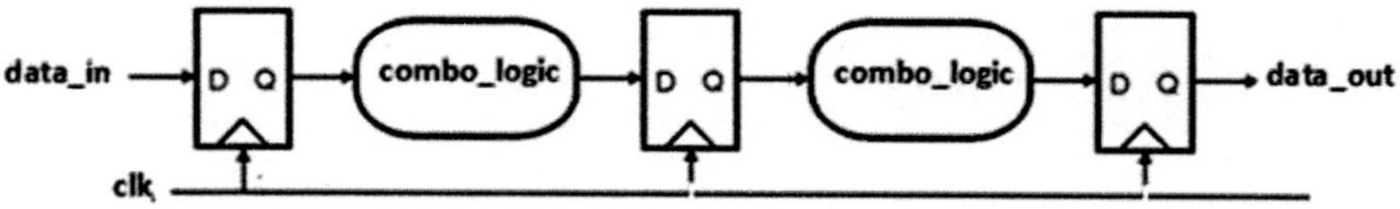

Fig. 15.14 Logic with register balancing

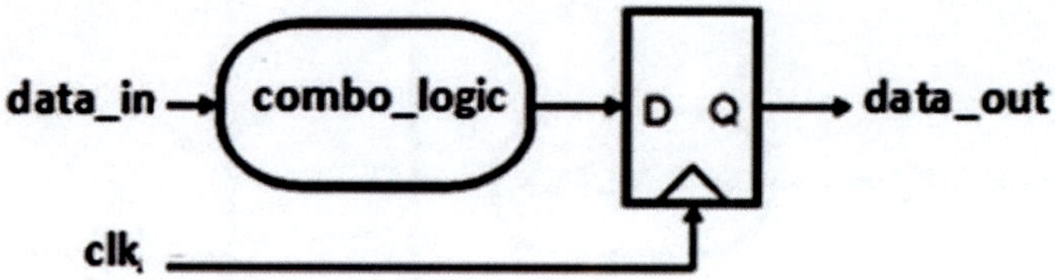

Fig. 15.15 Logic which has early arrival of data

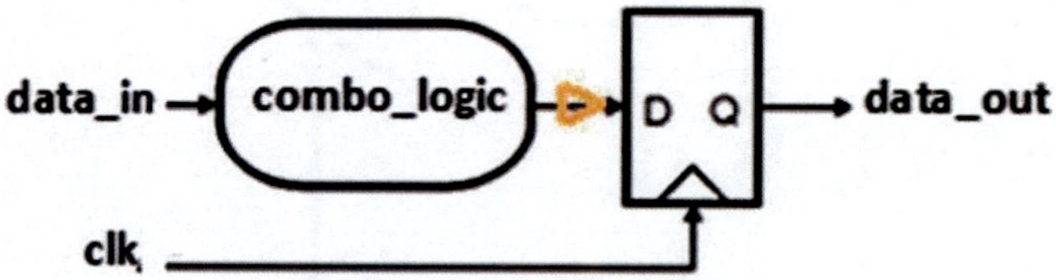

Fig. 15.16 Logic to fix the hold violation by using data buffers

15.5.2 Hold Violation Fix

Hold time violation occurs in the design if the data at the D input of flip-flop changes very fast. For example, consider the input to reg path shown in Fig. 15.15; if combinational logic delay is small, then there is chance of the data toggle during the hold window after arrival of active edge of the clock. This leads to the hold violation, and design will not meet the timing constraints.

To fix the hold violations, try to insert the buffers in the data path but take care that it should not have impact on the setup. Setup slack should be positive. The strategy is described in Fig. 15.16.

15.5.3 Timing Exceptions

There are two main timing exceptions and are named as false paths and multicycle paths; for more details, refer Chaps. 13 and 14. These timing exceptions need to be specified during the timing analysis.

15.6 Chapter Summary

The following are few of the points to conclude the chapter

1. STA is static timing analysis and is non-vectored approach to find the timing violations.

2. DTA is vectored approach and is dynamic timing analysis.
3. During STA, the objective is not to verify the functionality, but the major objective is to report the timing violations and fix them.
4. If setup or hold time is violated then design has timing violation.
5. The multicycle paths and false paths are the timing exceptions in the design and need to be specified.
6. The setup violations are due to late arrival of the data and can be fixed using the RTL, synthesis and architecture tweaks.
7. Hold violations are due to fast arrival of the data and can be fixed by having the data path analysis that is by inserting buffers in the data path.

Chapter 16
Physical Design

The physical design of the complex ASIC is time consuming and needs lot of attention to avoid the congestion and to improve the performance of the chip. During various stages, the common issues faced are due to

1. Congestion
2. Routing issues and routing delays
3. Issues during the distribution of the clock and clock skew
4. Issues due to the net delays and parasitic
5. Meeting of the chip-level constraints such as timing and maximum frequency
6. Issues due to noise and derate of the timing
7. Design rule check fails
8. LVS issues due to the routing.

All these issues need to be addressed during the physical design, and the strategies need to be developed to get the layout and GDSII with required timing and power.

The chapter discusses the physical design flow steps in details and strategies during the physical design.

16.1 Physical Design Flow

As discussed in Chap. 2, the gate-level netlist is available from the logic design flow. The netlist with chip constraints and required libraries are used as inputs during the physical design flow.

The physical design starts with the floor planning that is planning of the design mapping, and the goal is that there should not be congestion while routing of the design and the logic blocks or functional blocks should meet the aspect ratio. The better floor planning is required to have the better area, speed, and power and will

Physical design is also called as backend design, and the objective is to get the GDSII.

V. Taraate, *ASIC Design and Synthesis*,
https://doi.org/10.1007/978-981-33-4642-0_16

be useful to avoid the routing congestion. The power planning stage is used to plan for the power rings (VDD and VSS) and power straps depending on the power requirements.

After the power planning is done, the clock tree synthesis needs to be performed to balance the clock skew and to distribute the clock to the various functional blocks of the design. The clock tree can be H tree, X tree, balanced tree and discussed in this chapter.

The placement and routing are done to have the layout of the chip. The layout will have the routing delays, and many times the STA needs to be performed to find and fix the timing violations.

The layout of chip needs to be checked to verify the

(a) Foundry rules that is DRC
(b) LVS that is checking of the layout versus the schematic, and the intent is to verify the layout with the gate-level netlist.

If all the design rules are met and there are no any issues in the LVS, the team needs to perform the signoff STA. The reason being the after the layout, the design will not meet the required timing and frequency and may require the modification or tweaking at the various stages. The flow is iterative, and objective is to achieve the chip-level constraints.

After the signoff STA, the GDSII is generated. The GDSII stands for the Generic or Geometric Data Structure Information Interchange and describes the layout of the design with the connectivity.

The foundry uses the GDSII to manufacture the chip. It is also treated as tapeout delivered to foundry! (Fig. 16.1).

16.2 Foundation and Important Terms

Initially, the area estimation for the chip is unknown and with initial floorplan we can estimate the rough area utilization. The area optimization is especially important during the ASIC design, and the utilization is basically the percentage of the area that has been used in the chip. Important terms which we need to consider are following!

(a) ***Chip-Level Utilization***: It is the ratio of the area of standard cells, macros and the pad cells with respect to area of chip

$$\begin{aligned}&(\text{Area (Standard Cells)} + \text{Area (Macros)}\\&+ \text{Area (Pad Cells)})/\text{Area (chip)}\end{aligned}$$

(b) ***Floorplan Utilization***: It is defined as the ratio of the area of standard cells, macros, and the pad cells to the area of the chip minus the area of the sub floorplan

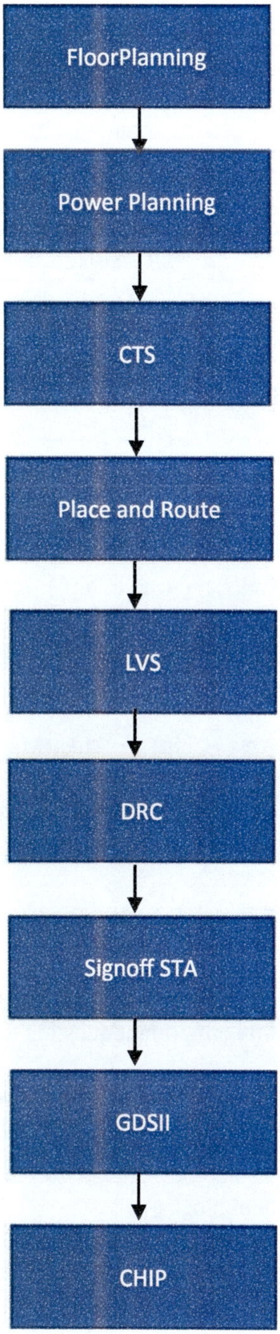

Fig. 16.1 Physical design flow

Table 16.1 List of various formats

Format	Description
DSPF: Detailed standard parasitic format	It contains the RC information of the nets
RSPF: Reduced standard parasitic format	It consists of the information about the RC delays in terms of a pi model
SDF: Standard delay format	The delays of the standard cells and net delays
SPEF: Standard parasitic exchange format	It contains the parasitic information
LEF: Library exchange format:	It contains the logic cell information
DEF: Design exchange format	It is used to describe the physical information of netlist
EDIF: Electronic design interchange format	It is used to describe schematics and layout
GDSII: Generic data structures library	This file is used by foundry to manufacture the ASIC
PDEF: Physical design exchange format	It consists of the clustering information

$$((\text{Area (Standard Cells)} + \text{Area (Macros)} + \text{Area (Pad Cells)})/(\text{Area (Chip)}) - \text{Area (sub floorplan)})$$

(c) ***Cell Row Utilization***: It is defined as the ratio of the area of the standard cells to the area of the chip minus the area of the macros and area of blockages

$$\text{Area (Standard Cells)}/(\text{Area (Chip)} - \text{Area (Macro)} - \text{Area (Region Blockages)})$$

Following are the important formats which we should know (Table 16.1).

16.3 Floor Planning and Power Planning

What is the information which is described by the netlist which is available after post-synthesis? The answer of this important question can be used to have the better floor plan!

The netlist which is available during logic design flow after the pre-layout STA is input to the floor planning tool consider as IC compiler. The netlist gives information about the

1. Various design and functional blocks
2. Macros
3. Memories
4. Interconnection between these block.

As the physical design team uses the netlist which is logical representation of the design with the goal to get the physical representation of the design. So, the design should have the better floor plan to get the design description. In simple words, the floor plan is the step in the physical design flow to get the physical description of the design.

What should be the strategies for the best floor plan?

To get the best floor plan, the objective of the team is to

1. Use of the minimum area
2. Strategy to have floor plan so the congestion can be very minimum
3. The delays due to routing can be minimized due to better floor plan.

So, during the floor planning we will perform the following important tasks

1. The chip area and size estimation
2. Strategy to arrange various functional blocks on the silicon
3. Strategy for the pin assignment
4. Planning for the IOs.

As per as ASICs are concern, the better floor planning is useful to improve the design performance by achieving the required timing and area. Most of the time during floor planning, we need to consider about the power planning and clock tree synthesis. But for easy understanding, the two different steps are documented as

1. Floor planning
2. Power planning
3. CTS.

Floorplan needs use of the important elements!

For any kind of ASIC, the important elements are

- **Standard cells**: For the specific technology node
- **IO cells**: To establish communication with the chip.
- **Macros**: Memories such as SRAM, DRAM (Fig. 16.2).

16.4 Power Planning

Depending on the power requirements, the VDD and VSS power rings are created across the floor plan.

Power Rings: Carries VDD and VSS around the chip

Power Stripes: Carries VDD and VSS from rings across the chip

Power Rails: Connect VDD and VSS to the standard cell VDD and VSS (Fig. 16.3).

Fig. 16.2 Floor planning the blocks relative to each other. *Image Courtesy* Andrew Kahng, UCSD

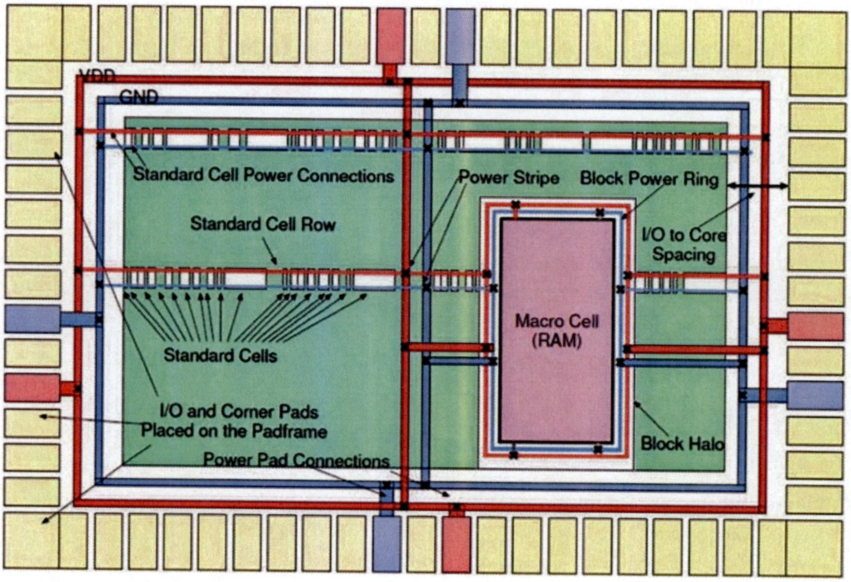

Fig. 16.3 Power planning [1]

16.5 Clock Tree Synthesis

Using the Tcl-based script, we will try to work on the clock tree synthesis.

Different clocking strategies are used to distribute the clock with the uniform clock skew, and few of the clock trees are

1. **Clock Tree**: to distribute the clock across the chip and to have the uniform skew (Fig. 16.4)
2. **H Tree**: To distribute the clock across the chip with uniform skew the another method is H tree synthesis (Fig. 16.5).

Use the following strategies while using the physical design tool

1. Check for the various tool-based optimization steps and enable them
2. Use the Tcl script to perform the CTS
3. Use the clock optimization. For example, using IC compiler, we can use

$$clock\_opt-effort\ high$$

4. Save the cell and report the following
   (a) Placement utilization
   (b) Quality of report (QOR).
   (c) Report timing that is setup and hold
5. Perform the post-optimization after CTS: For any kind of the setup and hold violations, perform the following
   (a) Fix the setup time
   (b) Clock optimization
   (c) Clock sizing
   (d) Hold time fix
   (e) Report the timing.

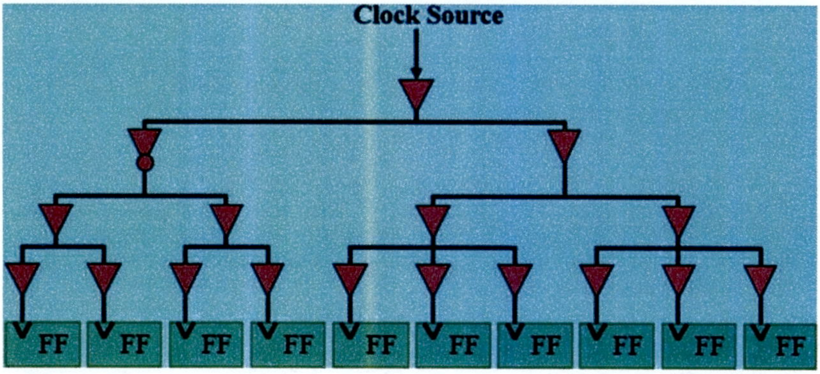

Fig. 16.4 Clock tree

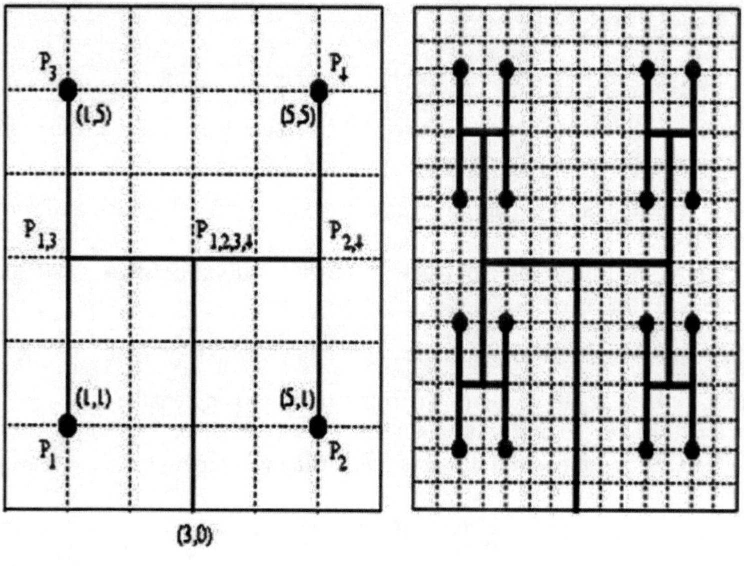

Fig. 16.5 H tree

16.6 Place and Route

During the floorplan, we do not have accurate information of the placing of the standard cells. During the placement, the physical locations of all the standard cells are defined. After the placement, the accurate estimation of capacitive load we can get at each standard cell.

The tool uses the placement algorithm to place the standard cells and set the space aside to have the better routing.

Following are important objectives during placement.

1. Use of the area constraints to have the location strategies so that the logic congestion can be minimum.
2. Have the less wire delays that is net delays so that the design performance can meet.
3. The placement should be useful to have better routing during the routing stage and there should not be cell overlaps.
4. Tool uses the algorithm which is using the timing and area constraints to have the global and detail placement.

Few commands which are useful during placement are

place_opt used to insert buffers to avoid the DRC violations.

Using the IC compiler, use the

set_buffer_opt_strategy -effort low to have the buffering strategy

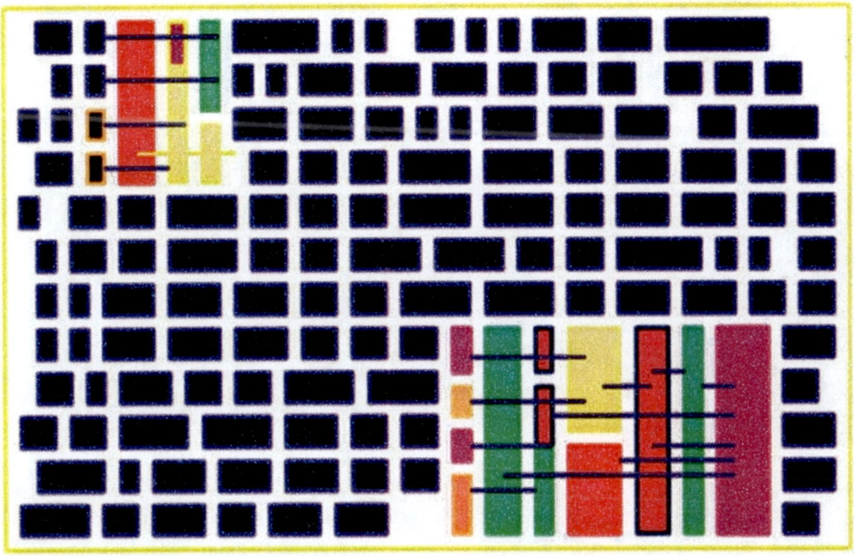

Fig. 16.6 Standard cells arranged on site rows. *Image Courtesy* Andrew Kahng, UCSD

There are other commands which are useful to tackle with the area and congestion, and these are

place_opt –congestion

place_opt -area_recovery -effort low

This can be used with various options as high, medium.

Used to report the placement utilization

report_placement_utilization

use **report_qor, report_timing** to report the quality of reports and timing after placement (Figs. 16.6 and 16.7).

16.7 Routing

After the placement stage, perform the global and detail routing and this is accomplished by using the physical design tool. Routing is basically connection of the different functional blocks of the design to establish the connectivity and communication (Fig. 16.8).

1. **Global routing**: It is used to perform the connections between all the blocks using small length nets and objective is to

   (a) minimize the length of interconnect
   (b) minimize the critical path
   (c) To determines the assignments for each interconnect

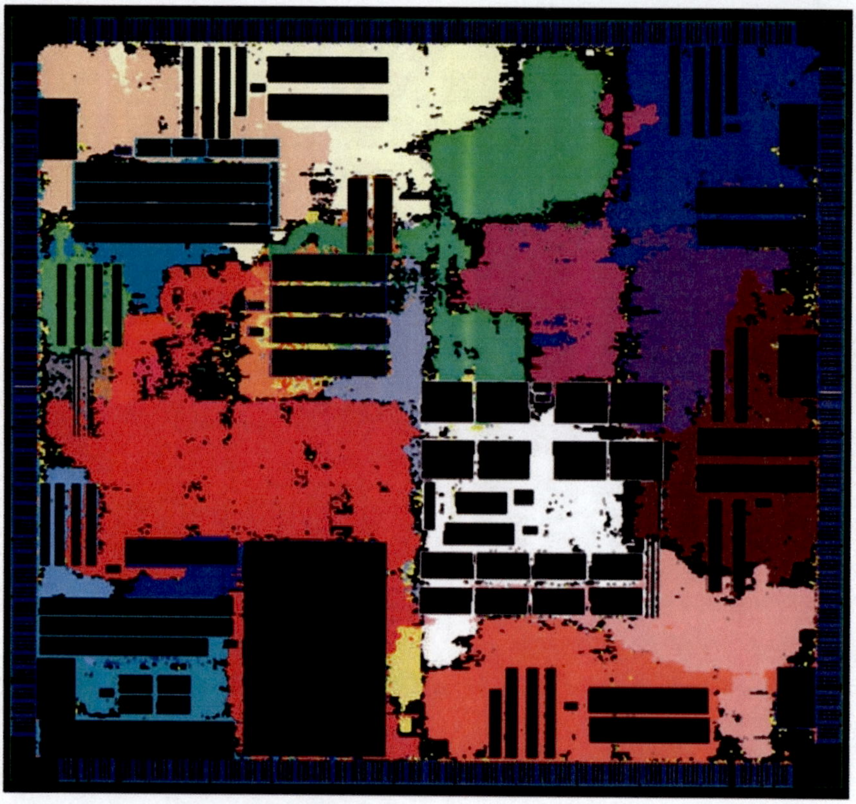

Fig. 16.7 Placed design. *Image courtesy* Andrew Kahng, UCSD

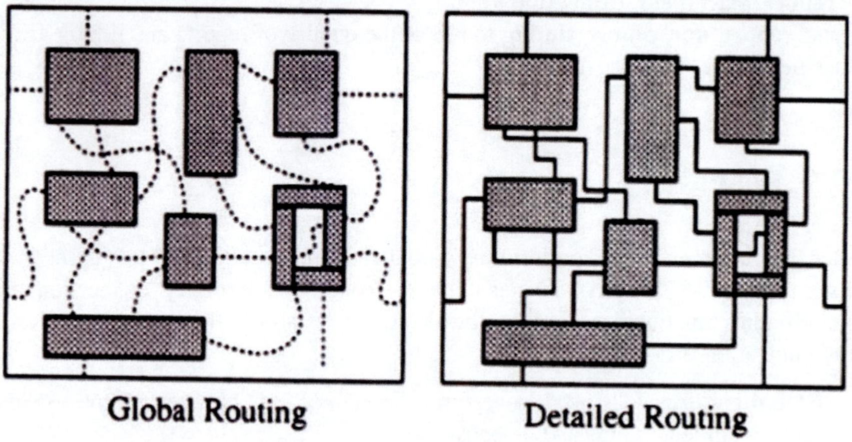

Fig. 16.8 Global and local routing

(d) Minimizing the congestion as it leads to the DRC issues.

The routing algorithm decides about all above parameters while performing the routing.

2. **Detailed Routing**: This is basically to establish the real connections between all the nets. This step is useful to create actual via and metal connections. The main objectives of detailed routing algorithm are to

   (a) Minimize the area
   (b) Minimize the wire length
   (c) Optimize delays in the critical paths.

During the final routing with the width, layer the exact location of the interconnection are decided.

Use the command **route_opt** to have the routing.

Now try to find the placement utilization using the command **report_placement_utilization**.

Report the timing and qor using the **report_timing, report_qor** and try to fix the timing issues.

To fix the issues, use the incremental route and optimize design commands (Fig. 16.9).

16.8 Back Annotation

The parasitic extraction that is finding out the R and C for each net is performed and use of the actual data to have the wire load model. Back annotate the information in the new wire load model to DC and to floorplan, P&R tool during the re-optimization of the design.

16.9 Signoff STA and Layout

Perform the timing analysis after the back annotation; if design does not meet the timing, then perform the resynthesis using the data available in the wire load model (back annotated data). For the smaller timing violations, use the re-optimization design strategies. Perform the above steps from the synthesis to P and R with the new wire load model data unless and until timing doesn't met.

Still the design has timing violation; then, reoptimize design and use the in-place optimization.

1. Optimization Strategies

Use the strategies like reoptimize design and in place optimization if the design does not meet the timing goals.

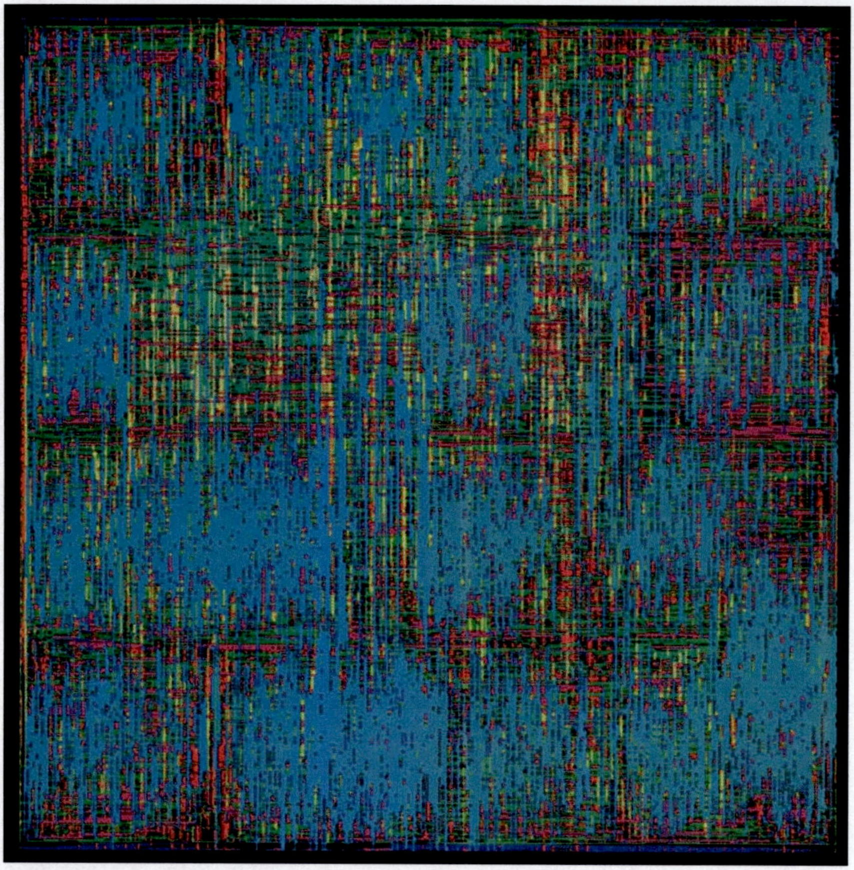

Fig. 16.9 Routed design. *Image Courtesy* Andrew Kahng, UCSD

The re-optimization uses the physical clustering and its information to meet the goals. If is similar like compile incremental but compile incremental works on the logical clusters.

After the re-optimization if the design doesn't meet the timing, then perform the in place optimization (IPO); it is the technique where the critical path cells can be swapped in other non-critical sub paths. The technique is useful to meet the constraints to have the layout with clean timing (Fig. 16.10).

For more details about the physical design and various algorithms, refer the Physical design and the IC compiler books.

Fig. 16.10 FRICO ASIC, 350 nm technology

16.10 Chapter Summary

Following are important points to conclude the chapter.

1. The netlist which is available during logic design flow after the pre-layout STA is input to the floor planning tool.
2. During the placement, the physical locations of all the standard cells are defined.
3. After the placement stage, perform the global and detail routing, and this is accomplished by using the physical design tool
4. The parasitic extraction that is finding out the R and C for each net is performed and use of the actual data to have the wire load models.
5. Signoff STA is performed after bank annotation phase as RC delays are available.

Reference

http://vlsibyjim.blogspot.com/2015/03/power-planning.html

Chapter 17
Case Study: Processor ASIC Implementation

The design of the processors from the functional specifications to the GDSII is time consuming as we need to tackle about the issues like data integrity for the multiple clock domains, processor configuration management and the architecture tweaks for the better performance. Even during the physical design stage, we need to consider about the floor plan strategies so that we can have room for the better routing. The major issue is due to large number of resources required to perform the floating-point operations and the timing requirements. If the block-level constraints are met, then also it is impossible to meet all the chip-level constraints during the signoff STA.

The chapter discusses about the various strategies during the RTL to GDSII of the processor designs.

17.1 Functional Understanding

If we consider the today's ASIC design, then the complexity of design is very high, and for the faster product launch, we may need to think about the use of processor IPs. There are many high density processor cores which are available in the market and used during the ASIC design cycle. The processor cores are used to perform various operations on the signed, unsigned, and floating-point numbers. They may have the parallelism and multistage pipelining features to boost the overall processor performance.

The objective of this chapter is to have discussion on how to transform the functional specifications into logic design and how to get the GDSII!

Processor cores are extensively used in the ASIC designs.

V. Taraate, *ASIC Design and Synthesis*,
https://doi.org/10.1007/978-981-33-4642-0_17

Let us consider the 32-bit processor which has following specifications.

1. It should perform the arithmetic operations such as addition, subtraction, multiplication, division, and modulus on signed, unsigned, and floating-point numbers.
2. It should perform the logical operations on 32-bit binary numbers.
3. It should perform the data transfer and branching operations.
4. It should perform the shifting and rotate operations.
5. The external interfaces can be
   (a) IO interfaces
   (b) Serial IO
   (c) High-speed interfaces.
6. It should have the internal memory storage of 64 KB
7. The processor should have the interrupt controller.
8. The processor has two clock domains and should use clk1 and clk2, respectively.

Consider that these specifications are extracted from the requirements and let us try to use these to have better architecture and micro-architecture.

17.2 Strategies During Architecture Design

Let us use the functional specification understanding to finalize the architecture of the processor. Use the following strategies to have the better architecture

1. **Multiple Clock domain groups**: The design is multiple clock domains, and we should have strategies to deploy the synchronizers. Following can be thought during the architecture design. Let us try to have the understanding of the functionality in these clock domains.

Clock domain 1: It is controlled by the clk1, and the functional blocks of this clock domain are

1. ALU
2. Internal memory
3. Interrupt controller
4. Pointers and counters
5. Serial IO
6. IO Interfaces.

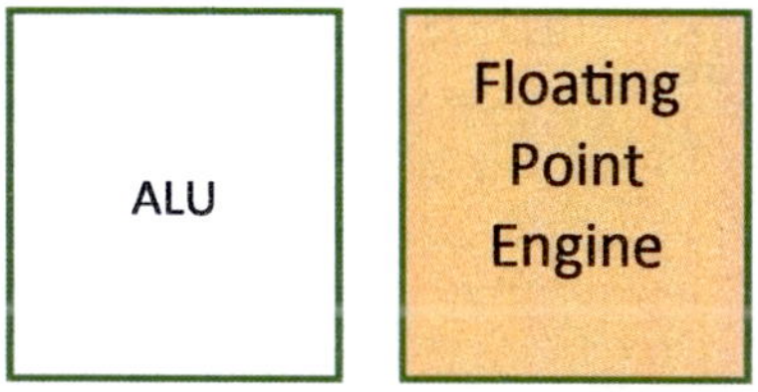

Fig. 17.1 Processor engine

In the architecture, clock domain 1 block is indicated by the white color boxes.

Clock domain 2: It is controlled by the clk2, and the functional blocks of this clock domain are

1. Floating-Point Unit.
2. High-speed interfaces.

In the architecture, clock domain 2 blocks are indicated by the yellow filled color.

2. **Processor Engine**

As stated in the specification extraction document, the processor core performs various operations on the signed and unsigned number and floating-point numbers, so the better strategy is to have the dedicated block for the general-purpose operations using ALU and floating-point operations (Fig. 17.1).

ALU: Performs the general-purpose operations on the signed, unsigned numbers

Floating-point Engine: Used to perform the operations on the floating-point numbers.

Then, let us try to understand the memory requirements and use the dedicated memory block of 64 KB may have partitioning according to the address ranges so that various functional units can perform the read and write operations.

3. **Memories**: To store the internal data, the processor needs to have the internal memory and can be shared between the general-purpose ALU and the floating-point engine. As we have the multiple clock domain design, then better way is to have the separate memory for the general-purpose processor and floating-point engine.
   With reference to the specification, the 64 KB memory is divided into two blocks of 16 KB and 48 KB, respectively (Fig. 17.2).
4. **High-Speed Interfaces**: To exchange the data from the external memory and IO after performing the floating-point operations, the architecture needs to have the high-speed interfaces. These high-speed interfaces need to be designed to have lower latency and minimum interconnect delays.

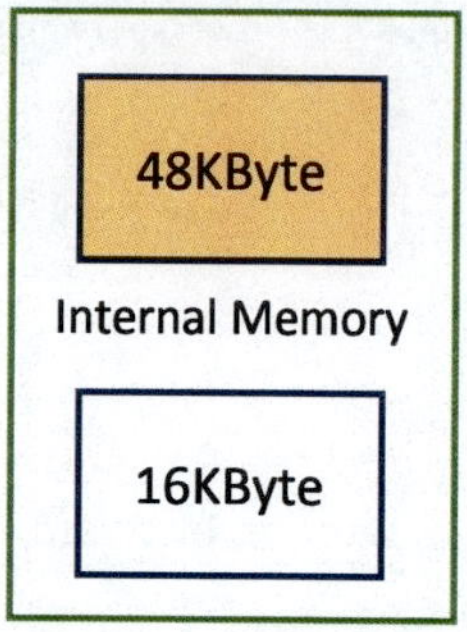

Fig. 17.2 Memory partitioning

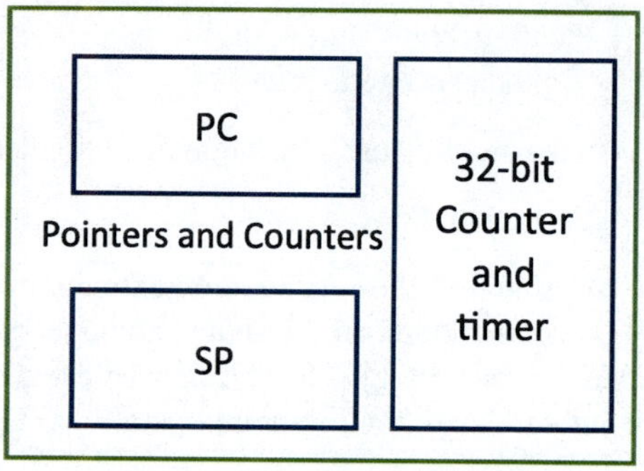

Fig. 17.3 Pointers and counters

5. **Pointers and Counters**: During the general-purpose processing of the data, the result may need to be stored in the reserved area of memory, so the design may need the stack pointer, and to fetch the instruction and the data from the external memory, the design needs to have the program counter. The stack and program counter are 16-bit for the architecture (Fig. 17.3).

The 32-bit counter and timer are used as dedicated timer and counter during the counting applications.

6. **IO and Communication Blocks**: To communicate with the external devices such as serial and parallel, the processor architecture should have the dedicated blocks. These blocks are

   - **IO Interfaces:** For the 32-bit data transfer dedicated high-speed IOs to exchange the 32-bit of the data between the IO devices and processor.
   - **Serial IO:** The serial devices can communicate with the general-purpose processors using the serial IOs.

7. **Interrupt Controller**: The architecture provides the dedicated block to process the edge- and level-sensitive interrupts. The interrupt controller can halt the current execution for the valid interrupt (Fig. 17.4).

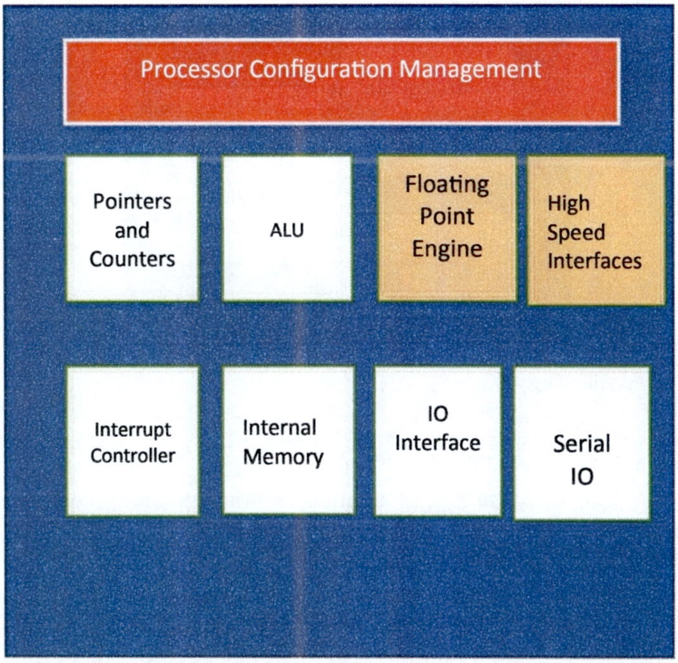

Fig. 17.4 Processor architecture

17.3 Micro-architecture Strategies

As discussed in Chap. 99, the micro-architecture is the sub-block level; representation and the following strategies we can use to have the better micro-architecture

1. Try to have the rough initial estimation for the functional density of each block.
2. Depending on the functional requirements, try to have the sub-block-level representation. For example, the ALU we can indicate as the (Fig. 17.5)
3. For each functional block, sketch and document the sub-block-level representation.
4. Identify the interfaces for each sub-block and try to document them with timing information.
5. If the architecture demands use of the IPs, then the functional and interfaces can be documented with the timing information.
6. Multiple clock domain functional units try to describe as separate groups.

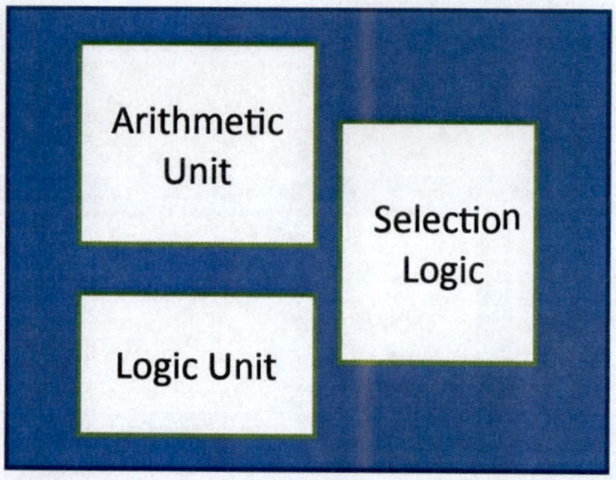

Fig. 17.5 ALU partitioning

Let us try to have the understanding of the micro-architecture for the 32-bit ALU. The ALU performs the operations as addition, subtraction, multiplication, division, and modulus using the arithmetic unit and logical operations such as and, or, xor, not using the logic unit. So, the micro-architecture has two sub-blocks that is arithmetic unit and logic unit. The selection logic at output is used to select from one of the outputs. The sub-block-level representation of ALU is shown in Fig. 17.6.

The interface information with the associated blocks of the processor is documented in Table 17.1.

The tweaks recommended during the micro-architecture design that the external interfaces need to have registered inputs and outputs so the following micro-architecture for the ALU is better option (Fig. 17.7).

The micro-architecture for the ALU has separate data and control path and has registered inputs and registered outputs.

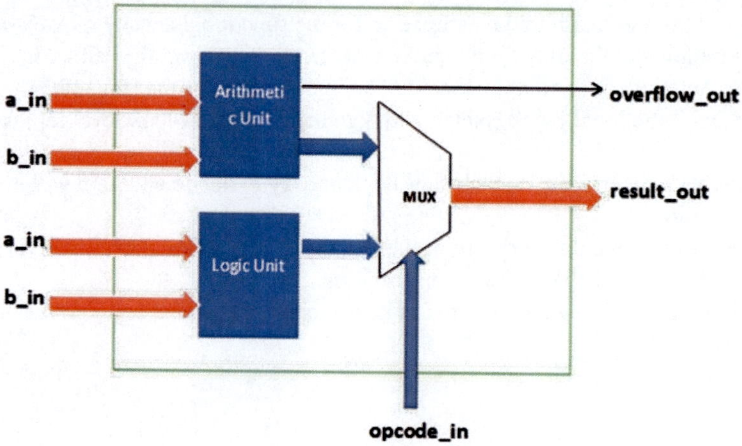

Fig. 17.6 Micro-architecture for ALU

Table 17.1 IO interfaces for ALU

Name of signal	Direction	Width	Description
a_in	Input	32-bit	Input to the ALU
b_in	Input	32-bit	Input to the ALU
opcode_in	Input	8-bit	Input to ALU to carry opcode
result_out	Output	72-bit	To carry result and flag information
overflow_out	Output	1-bit	To indicate the result overflow that is if result is greater than 64-bits, then flag = 1

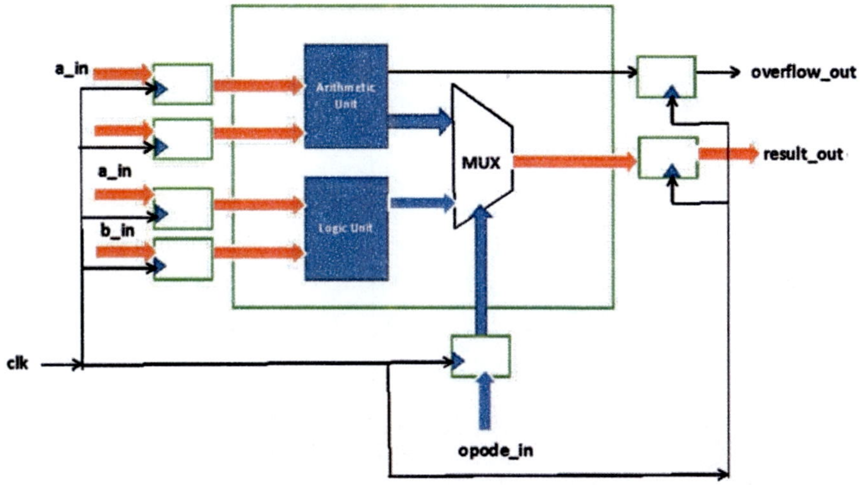

Fig. 17.7 Micro-architecture tweaks

By considering the above strategies, the micro-architecture is created for other functional blocks, and the sub-block-level representation is shown in Fig. 17.8.

17.4 Strategies During RTL Design and Verification

Following strategies were used during the RTL design and verification of the processor

1. Partitioning of the block-level design to improve the overall area and speed of the design.
2. Use of the case constructs instead of if-else to avoid the priority logic.
3. Have the separate FSM controllers to have the better timing control.
4. Have the separate modules for the control and data path synchronizers.

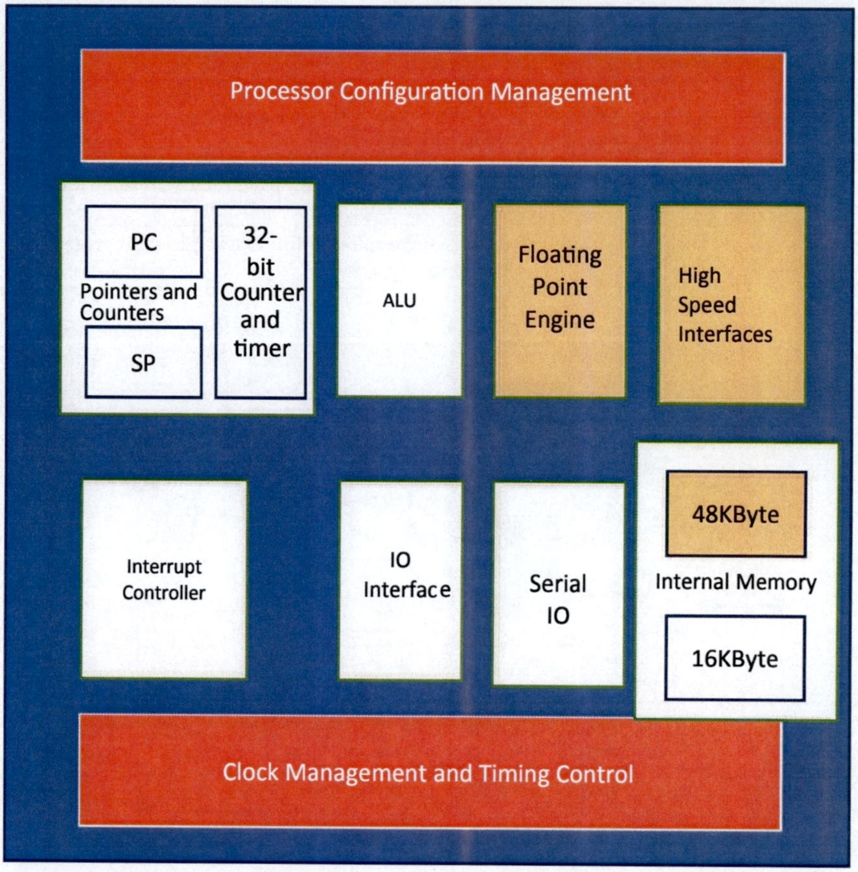

Fig. 17.8 Processor micro-architecture

5. Having the use of the resource sharing and pipelining using synthesizable constructs.

During the **RTL verification,** the following strategies were used

1. Have the better block-level and top-level verification plan and architecture.
2. Documenting the corner cases and the test cases for the block-level designs and for the top-level design. For example, the multiplication by 0, division by 0, overflow and the flag generation checks.
3. Use of automatic testbenches during the verification.
4. Checking for the coverage: Functional, code, toggle, etc.
5. Monitoring and documenting the results for the block- and top-level design.

17.5 The Sample Script Used During Synthesis

The sample script can be used to constrain the design for operating frequency of 500 MHz

```
/* set the clock */
set clock clk
/* set clock period */
set clock_period 2
/* set the latency */
set latency 0.05
/* set clock skew */
set early_clock_skew [expr $ clock_period/10.0]
set late_clock_skew [expr $ clock_period/20.0]

/* set clock transition */
set clock_transition [expr $ clock_period/100.0]
/* set the external delay */
Set external_delay [expr $ clock_period*0.4]
/* define the clock uncertainty*/
set_clock_uncertainty –setup $ early_clock_skew
set_clock_uncertainty –hold$ late_clock_skew

Name the above script as clock.src, and Source the above script

/* report clock and timing*/
dc_shell> report_timing
dc_shell> report_clock
dc_shell> report_timing
dc_shell> report_constraints –all_violations
```

17.6 Synthesis Issues and Fixes

Following are few of the scenarios which we can understand the fix during the synthesis and optimization.

1. ***Area Optimization***: The Synopsys DC will not be able to optimize the hierarchical designs, and hence, for the ALU design, the area is not optimized.
   Solution: Try to tweak the micro-architecture and in turn the RTL by having strategy to perform the logical operations using the arithmetic resources. By using this strategy, the single block (ALU) is suitable and useful to perform the arithmetic and logical operations.

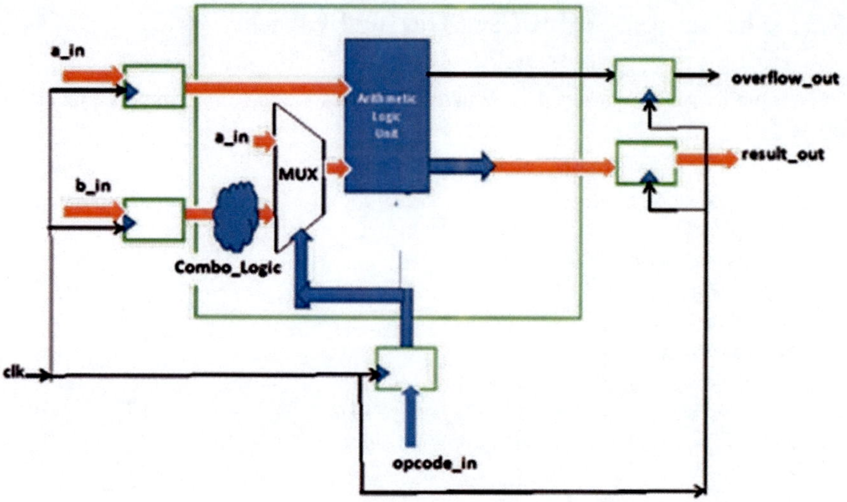

Fig. 17.9 Micro-architecture tweak to fix late arrival signals

2. ***Late arrival inputs opcode***: From the decoding engine, the opcode is arriving late and impacts on the timing. The setup time violates for the micro-architecture (Fig. 17.7)
   Solution: Use the strategy to push the common resources toward output side (ALU pushed toward output side) and use the combinational logic at the input side to improve the data path synthesis. This tweak at the micro-architecture level is useful to eliminate the setup time violation during the block-level synthesis.

As shown in Fig. 17.9, the opcode_in is pushed at the input side so that we can have the clean reg to reg path to eliminate the setup violation.

17.7 Pre-layout STA Issues

The following are few of the issues which need to understand and fix during the pre-layout STA.

1. ***Issue in the General-purpose processor timing***: Able to meet the timing for the general-purpose processor but still chance of the performance improvement as setup slack is very high.
   Solution: Improve the design performance using the pipelined architecture. Tweak the micro-architecture using the strategy shown in Fig. 17.10
   This strategy increases the area and introduces latency of few clocks to get the result but improves the design performance.
2. ***Floating-Point Engine Constraint violation***: During the block-level synthesis of the processor, the block-level constraints are met but the real scenario which

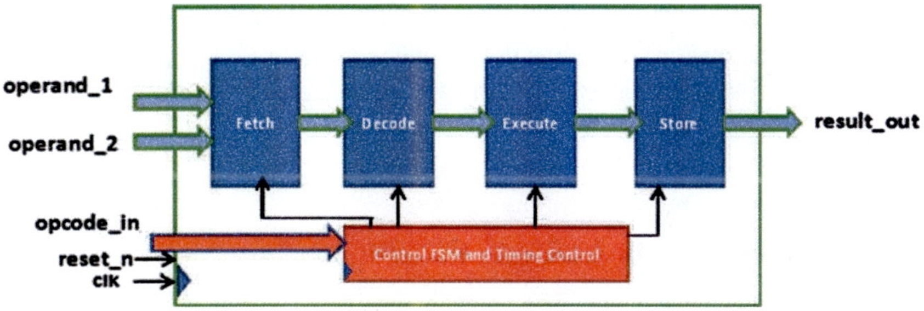

Fig. 17.10 Pipelining stage architecture

I faced during the top-level synthesis for the operating frequency of 500 MHz. The general-purpose processor timing was met that is setup slack was positive, but for the floating-point engine, the setup slack was negative and had the timing violation in the design.

Solution: The floating-point engine consists of many multipliers to perform the floating-point operations and was consuming the longest delay between the register. The critical path for the design was almost around 3.3 ns, and to eliminate the setup time violations, the following strategies were used:

1. Register balancing and optimization
2. Splitting of the combinational design by introducing the latency of one clock.
3. Tweaking of the RTL using the logic duplication.

The strategies were helpful to have the speed improvement and to meet the setup slack.

3. ***Timing Exceptions***: The floating-point engine design has lot of timing issues and not able to eliminate them during synthesis and optimization.

   Solution: The reason is it uses the large density multipliers, and hence, it is obvious that the design has the timing exceptions like

   - Multicycle path
   - False path.

   Notify these exceptions and then re-compile with the optimization goal.

17.8 Physical Design Issues

Consider that after the P and R design has the timing violations.

Solution: Use the strategies like reoptimize design and in place optimization if the design does not meet the timing goals.

The re-optimization uses the physical clustering and its information to meet the desired goals. It is similar like compile incremental but compile incremental works on the logical clusters.

After the re-optimization, if the design doesn't meet the timing then perform the in place optimization (IPO), it is the technique where the critical path cells can be swapped in other non-critical sub-paths. The technique is useful to meet the constraints and to have the layout with clean timing.

17.9 Chapter Summary

Following are the important points to conclude the chapter

1. If the architecture demands use of the IPs, then the functional and interfaces can be documented with the timing information.
2. For multiple clock domain designs try to have separate clock groups.
3. If design uses the large density multipliers, then it is obvious that the design has the timing exceptions like multicycle path.
4. Few issues during the pre-layout STA are block-level functionality timing meeting but not at top level.
5. To fix the setup violation, use the register balancing and optimization, splitting of the combinational design by introducing the latency of one clock, tweaking of the RTL using the logic duplication.

Chapter 18
Programmable ASIC

Modern ASIC designs are very complex and can consist of the million or billion gates. Before ASIC goes through the manufacturing process, it is essential to prototype the design to check for the functional correctness of the design. Even at the filed how the design behaves that need to be verified at the system level, hence it is essential to understand about the FPGA that is programmable ASIC flow and the ASIC synthesis. The following sections are useful to understand the FPGA synthesis and design using FPGA.

18.1 Programmable ASIC

To have the lowest NRE cost and to prototype the ASICs, the multiple FPGA architectures are useful. The prototype team uses the complex FPGAs to test the functionality and the connectivity of the different design blocks. The FPGA layout at fabric level is shown in Fig. 18.1, and it has various functional blocks such as

---

FPGA is programmable ASIC, and multiple FPGAs are used to prototype the complex ASICs.

V. Taraate, *ASIC Design and Synthesis*,
https://doi.org/10.1007/978-981-33-4642-0_18

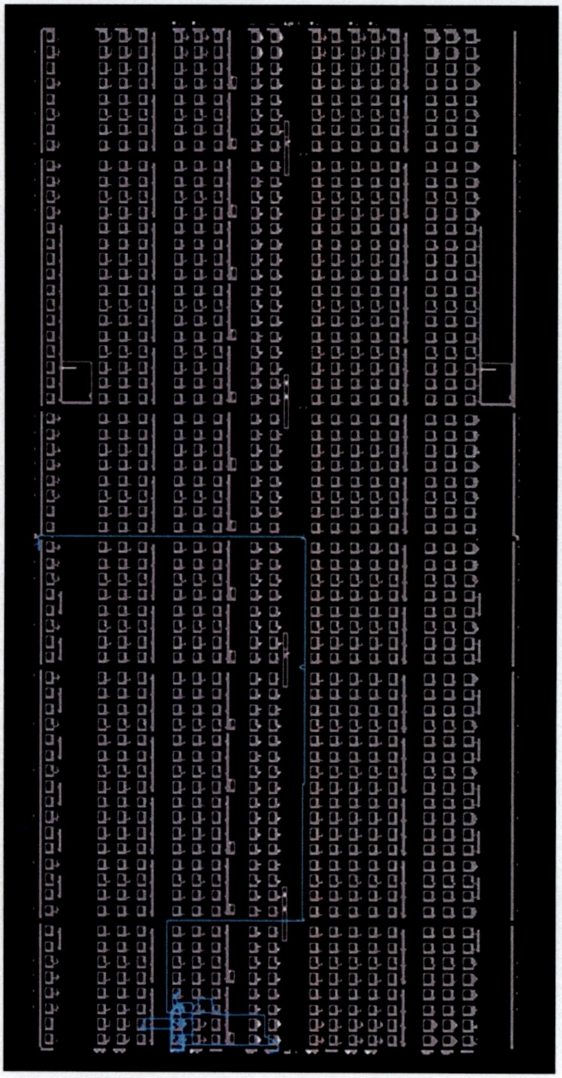

Fig. 18.1 FPGA fabric and layout

1. CLB (slice registers, LUTs, Mux)
2. Clock managers
3. IOBs
4. Multipliers
5. DSP blocks
6. Block RAMs.

Even the modern FPGAs have the processor cores, high-speed interfaces, and the memory controllers to access and process the larger amount of the data.

18.2 Design Flow

FPGA design flow can be also treated as programmable ASIC flow and described in Fig. 18.2.

The important steps are

1. Design planning
2. RTL design and verification
3. Synthesis
4. Design implementation: It consists of the following steps
    (a) Logic functionality mapping
    (b) Place and rout
    (c) SDF-based verification
    (d) Signoff STA.
5. Device programming.

Few of the important FPGA blocks are shown in Fig. 18.3: FPGA architecture. The modern FPGA architecture is complex and consists of few of the important blocks as follows:

1. **Configurable Logic Block (CLB)**: The array of CLBs is useful to map the logic using the LUTs, slice registers, and multiplexers to have the desired functionality.
2. **IO Blocks**: At the periphery of the FPGAs, the IO blocks are used to communicate with the external world and the logic on the FPGA fabric.
3. **Switch Boxes**: The switch boxes are used to establish connectivity of the different CLBs.
4. **DSP Blocks**: The dedicated blocks used as programmable resource for the complex DSP functionality.
5. **Multipliers**: The dedicated multipliers on fabric to perform the multiplication.
6. **Processor block**: The processor core which can be configured to perform the processing of the data.

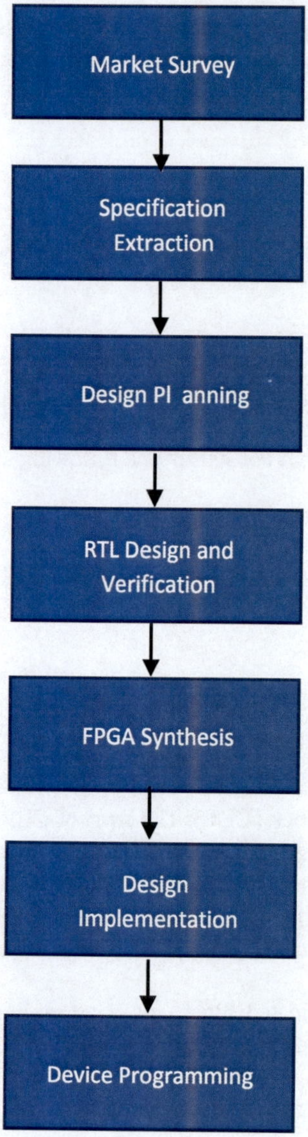

Fig. 18.2 FPGA design flow

18.3 Modern FPGA Fabric and Elements

The modern FPGAs have the complex architectures and useful to prototype the ASICs. Single FPGA may or may not accommodate the entire ASIC functionality so the prototype team needs to plan the ASIC prototype using the multiple FPGAs.

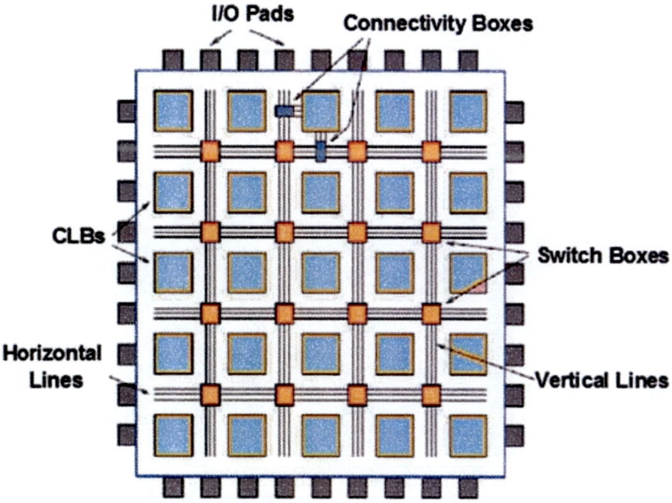

Fig. 18.3 FPGA architecture

As discussed in above section, the FPGA consists of the programmable blocks. The section discusses the programmable blocks for the modern FPGAs. The important FPGA blocks residing on the FPGA fabric are shown in Fig. 18.4.

Few of the blocks useful during prototype are discussed in this section

1. **Configurable Logic Block (CLB)**: The CLB consists of the slice register and the LUTs with associated logic such as adder with carry chain and multiplexers. The CLB for the Xilinx FPGA which consists of the six inputs LUTs and slice registers with the associated logic is shown in Fig. 18.5.

In the simple way, the CLB logic can be easily interpreted as the logic block which can be programmed to have the combinational and sequential design outputs. For example, consider (Example 1).

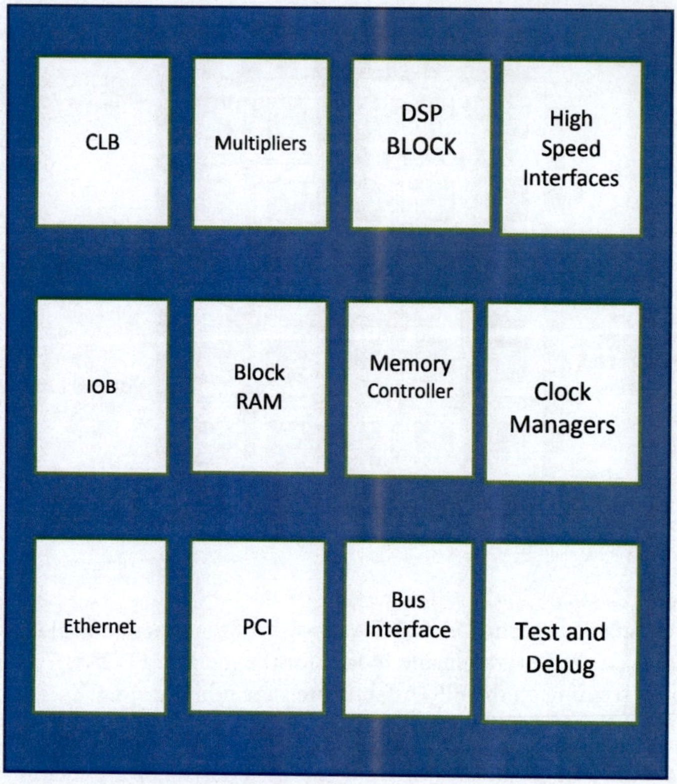

Fig. 18.4 Important FPGA vendor-specific blocks

Example 1 RTL design using Verilog

```
//////////////////////////////////////////////////
module fpga_design(input clk, a_in,b_in,sel_in, output  q2_out);
reg q1_out;
  always @ (posedge  clk)
  begin
  q1_out <= q_out;
  end
  assign q_out = a_in ^ b_in;
  assign q2_out = (sel_in) ? q1_out : q_out;
endmodule

//////////////////////////////////////////////////
```

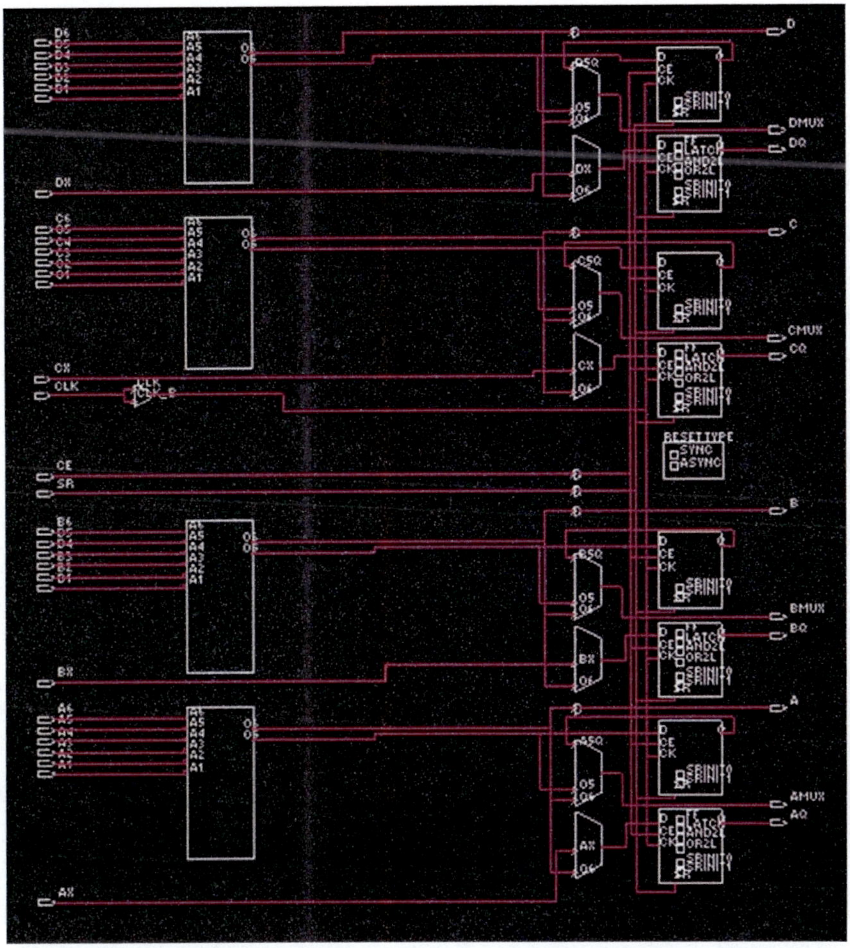

Fig. 18.5 CLB architecture

The synthesis if the Example 1: RTL design using Verilog infers the programmable logic residing within the CLB using the slice register and LUT (Fig. 18.6).

2. **Input/Output Block (IO Block)**: The IOs are around the periphery of the FPGA and the IO blocks are used to establish communication between the logic within the FPGA with the external world.

The IO block structure which is configured as an input to the FPGA is shown in Fig. 18.7, and as shown the data flow is from the PAD to the FPGA logic.

IO block which is configured as an output is shown in Fig. 18.8, and the data flows from the FPGA logic blocks to the PAD. Even IOs can be configured as bidirectional for the data transfer between the FPGA logic and external world.

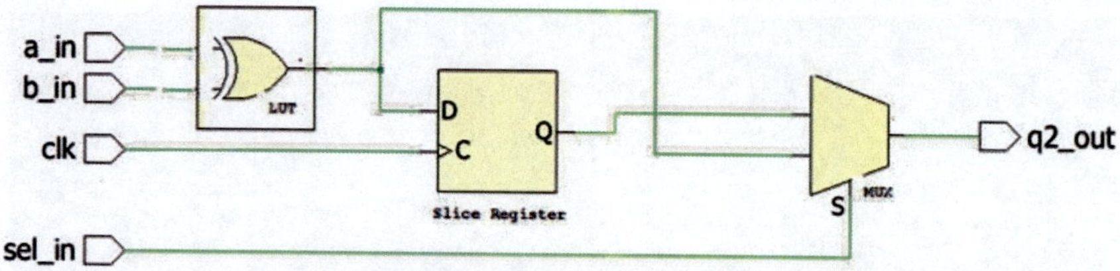

Fig. 18.6 Use of LUTs and slice register during synthesis

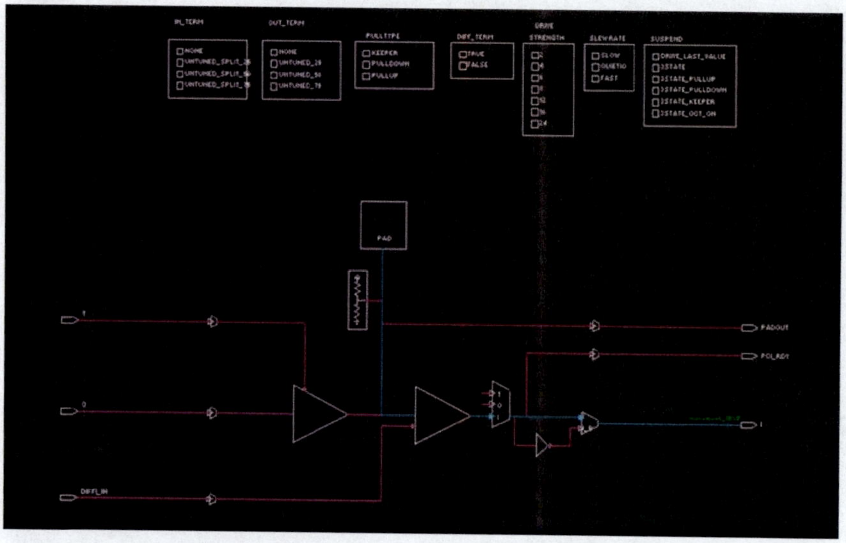

Fig. 18.7 Input block to the FPGA

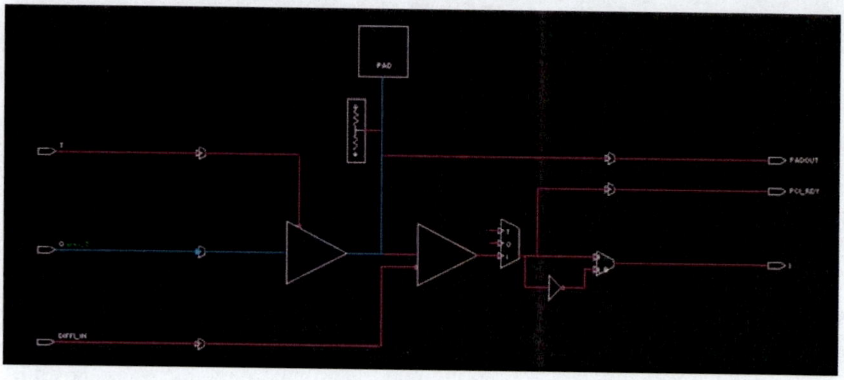

Fig. 18.8 Output block

3. **Block RAM**: The FPGAs have the block memories to store the data and called as block RAM.
4. **Multipliers**: The FPGA architecture is vendor-specific, and few of the FPGAs have the dedicated multipliers.
5. **DSP Blocks**: The complex DSP functionality can be implemented using the DSP blocks, and the multiple DSP blocks reside on the FPGA fabric.
6. **Clock Managers**: The clock managers are used to distribute the clock within the FPGA fabric with the uniform delay.
7. **Controllers**: The modern FPGAs have the high-speed memory controllers and processors.
8. **Interfaces**: Modern FPGA architectures have the provision to establish the connectivity using the high-speed interfaces.

18.4 RTL Design and Verification

The RTL design for the complex logic needs the better architecture and partitioning of the design. During the RTL design phase, the following strategies can be helpful

1. Have the detail understanding of the architecture and micro-architecture.
2. Have the partitioning strategies which can be useful to code the functionality using modular approach.
3. Use the design guidelines and have better understanding of FPGA architecture.
4. Have better understanding of the ASIC to FPGA conversions such as gated clocks, designs with clock enables.
5. Deploy the synchronizers for the multiple clock-level designs.
6. Use the separate modules for the FSM controllers and work on the strategies for the better data and control path synthesis.

Consider the design which uses the gray counter, the RTL description is shown in the Example 2 and the RTL schematic is shown in Fig. 18.9.

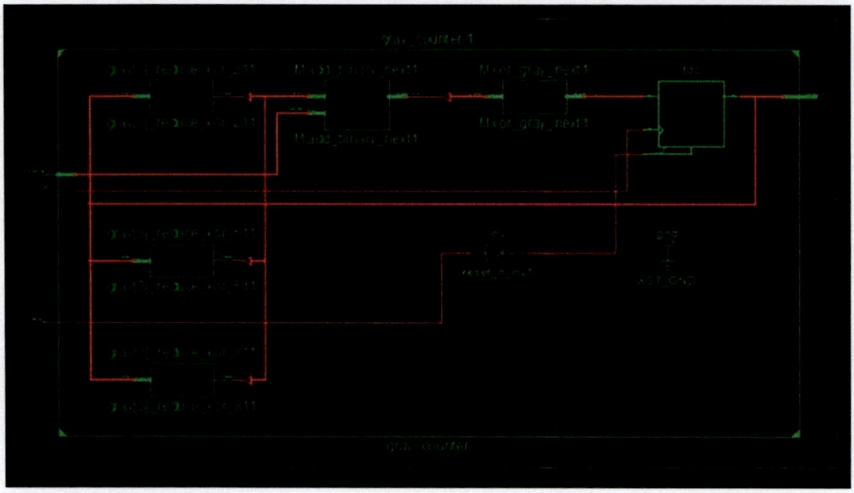

Fig. 18.9 RTL schematic for 4-bit gray counter

Example 2 RTL description for 4-bit gray counter

```
//////////////////////////////////////////////

module gray_counter (parameter data_size =4)
 (input clk;
input reset_n;
input increment;
output reg [data_size-1:0] gray;
);
parameter data_size =4;
reg [data_size-1:0] gray_next, binary_next, binary;
integer m;
always@(posedge clk or negedge reset_n)
if (~reset_n)
gray <= 4'b0000;
else
```

```
gray <= gray_next;

always@(*)
begin
for (m=0; m < data_size; m=m+1)
begin
        binary[m] =^ (gray >> m);
        binary_next = binary + increment;
        gray_next = (binary_next >>1) ^ binary_next;
end
end
endmodule
//////////////////////////////////////////////////
```

The technology schematic for the design is shown in Fig. 18.10.

The testbench for the gray counter is described using the non-synthesizable Verilog constructs and is shown in Example 3.

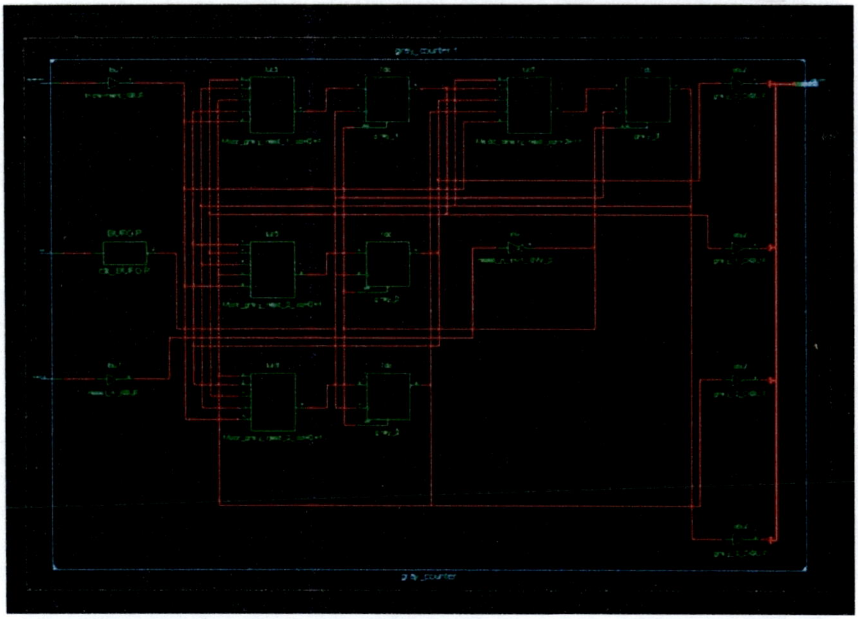

Fig. 18.10 Technology schematic for 4-bit gray counter

Example 3 Testbench of 4-bit gray counter

```
/////////////////////////////////////////////////


module test_gray;

        // Inputs
        reg clk;
        reg increment;
        reg reset_n;

        // Outputs
        wire [3:0] gray;

        // Instantiate the Unit Under Test (UUT)
        gray_counter uut (
                .clk(clk),
                .increment(increment),
                .reset_n(reset_n),
                .gray(gray)
        );
always #10 clk=~clk;
        initial begin
                // Initialize Inputs
                clk = 0;
                increment = 0;
                reset_n = 0;

                // Wait 100 ns for global reset to finish
                #100;

                increment =1;
                reset_n=1;

        end

endmodule

/////////////////////////////////////////////////
```

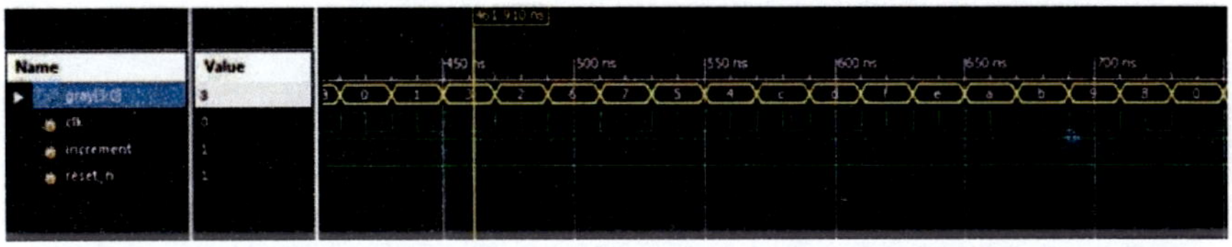

Fig. 18.11 Simulation result of 4-bit gray counter

The simulation waveform is shown in Fig. 18.11 as the counter generates the gray code on every positive edge of the clock during the inactive reset.

18.5 FPGA Synthesis

The FPGA uses the CLBs, IOBs, BRAMs, and other blocks to implement the design functionality.

As discussed in Chap. 1 the ASIC uses the standard cells and macros to infer the logic.

For moderate gate count designs which uses few logic gates, the RTL schematic may look similar but practically the FPGA synthesis and ASIC synthesis differ a lot due to the use of the resources.

18.5.1 Arithmetic Operators and Synthesis

Let us consider the use of the arithmetic operators such as + (Addition), − (Subtraction), * (Multiplication),/(Division) and %(modulus) shown in the RTL then for ASIC or FPGA synthesis the logic is inferred using the standard cells and LUTs, respectively.

The RTL for the arithmetic unit is described in the Example 4.

Example 4 Use of arithmetic operators

```
////////////////////////////////////////////////////////////////////////////


module arithmetic_operator_synthesis ( input [1:0] a_in, b_in,
            output reg [2:0] y1_out,
            output reg [1:0] y2_out,
            output reg [3:0] y3_out,
            output reg [1:0] y4_out,
            output reg [1:0] y5_out );
```

```
always @ *
begin

  y1_out = a_in + b_in;          // addition operator
  y2_out = a_in -b_in;           // Subtraction operator
  y3_out = a_in * b_in;          // multiplication operator
  y4_out = a_in /b_in;           // division operator
  y5_out = a_in % b_in;          // modulus operator

 end
endmodule
///////////////////////////////////////////////////////////////////////////
```

The RTL schematic for the design is shown in Fig. 18.12 and has the arithmetic resources to perform the desired operation. All the operations are performed in parallel to generate the parallel outputs.

18.5.2 Relational Operator and Synthesis

If we need to have combo logic to compute the less than and greater than, then we will use the relational operators.

The RTL described in the Example 5 uses the, < (less than), ≤ (less than equal to), > (greater than), ≥ (greater than equal to) operator, and the RTL: Schematic is shown in Fig. 18.13.

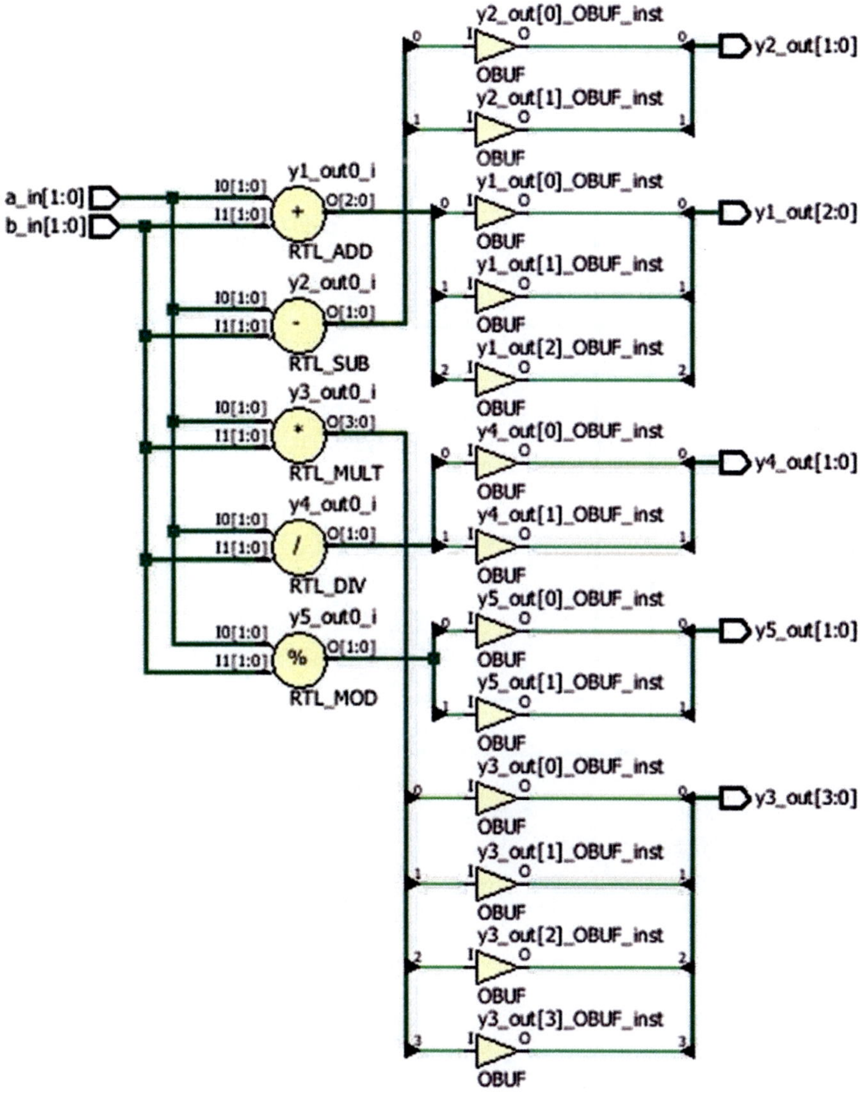

Fig. 18.12 Synthesis result for Example 4

Example 5 Use of the relational operators

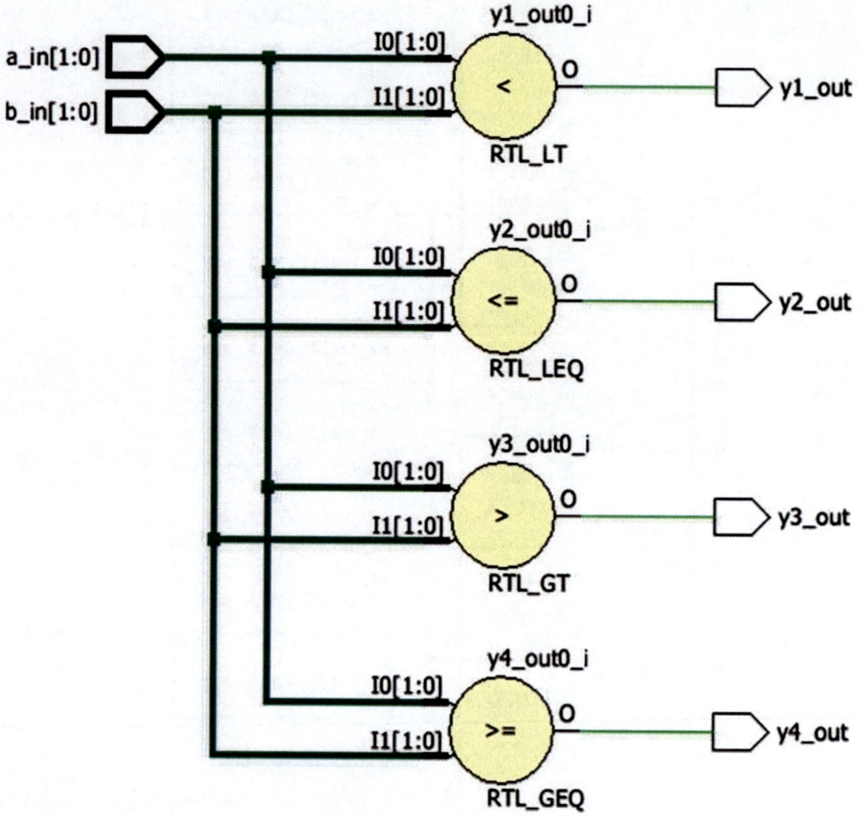

Fig. 18.13 RTL synthesis for the Example 5

```
//////////////////////////////////////////////////////////////////////////////////

module relational_operator (input [1:0] a_in, b_in,
          output reg y1_out,
          output reg y2_out,
          output reg y3_out,
          output reg y4_out
          );

 always @ *
 begin
  y1_out = a_in < b_in;         // less than operator
 y2_out = a_in <= b_in;         // less than equal to operator
 y3_out = a_in > b_in;          // greater than operator
 y4_out = a_in >= b_in;         // greater than and equal to operator
 end
endmodule
//////////////////////////////////////////////////////////////////////////////////
```

18.5.3 Equality Operator Synthesis

Most of the time we need to compare the strings during ASIC or FPGA designs, and in such scenarios we can use the equality operators.

The RTL description shown in the Example 6 uses the ==(equality),! = (inequality) operator a,nd the RTL schematic is shown in Fig. 18.14

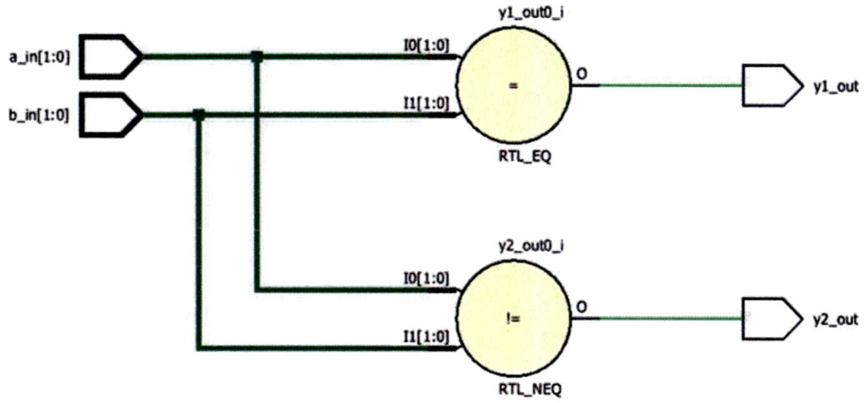

Fig. 18.14 RTL schematic for Example 6

Example 6 Use of equality operator

```
////////////////////////////////////////////////////////////////////////////
module equality_operator
        (

            input [1:0] a_in, b_in,
            output reg  y1_out,
            output reg  y2_out
                    );

  always @ *
  begin
    y1_out = (a_in == b_in);              //equality operator
    y2_out = (a_in != b_in);              // inequality operator
  end
endmodule
////////////////////////////////////////////////////////////////////////////
```

18.6 Design at Fabric Level

For the FPGA, the physical design consists of the following steps.

1. **Design Implementation**:
   (a) Logic functionality mapping
   (b) Place and rout
   (c) SDF-based verification
   (d) Signoff STA.
2. **Device Programming**

For the complex designs, the following issues need to be fixed

1. ***The trimming of the larger number of blocks during the place and route stage***: Try to check for the redundant logic and tweak the RTL.
2. ***The timing exceptions due to multicycle and false paths in design***: Specify the timing exceptions during STA.
3. ***The design is not fitting on the fabric and resource requirement is more that 100%***: Try to optimize for the area by enabling the tool directives. If still the design doesn't fit on the FPGA fabric, then use the area optimization techniques that are, try to tweak the RTL and architecture of the design.
4. ***Timing fails***: Try to optimize with the timing goals or use the performance improvement techniques such as register balancing and optimization.

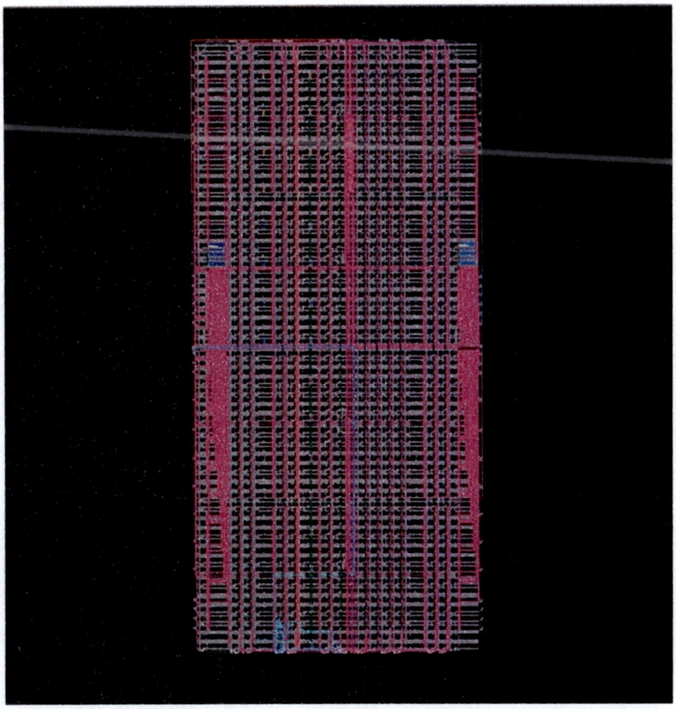

Fig. 18.15 FPGA fabric

The logic residing on the FPGA fabric which we can treat as the layout of FPGA is shown in Fig. 18.15.

The placement and routing are performed to actually place the logic on the FPGA fabric, and the design snapshot with various clusters is shown in Fig. 18.15. The routing is shown by the green color wires, and the length of the wires to route the design is chosen using the routing algorithm to have the least delays (Fig. 18.16).

After the final routing for the design, the SDF-based verification is carried out, and the signoff STA is performed. For any timing issues, the design needs to optimize or it is essential to perform the tweaks at the RTL and architecture level. The flow is iterative and time consuming for the complex designs.

After meeting all the timing goals, the FPGA device is programmed, and system's tests and verification at the device level can be performed.

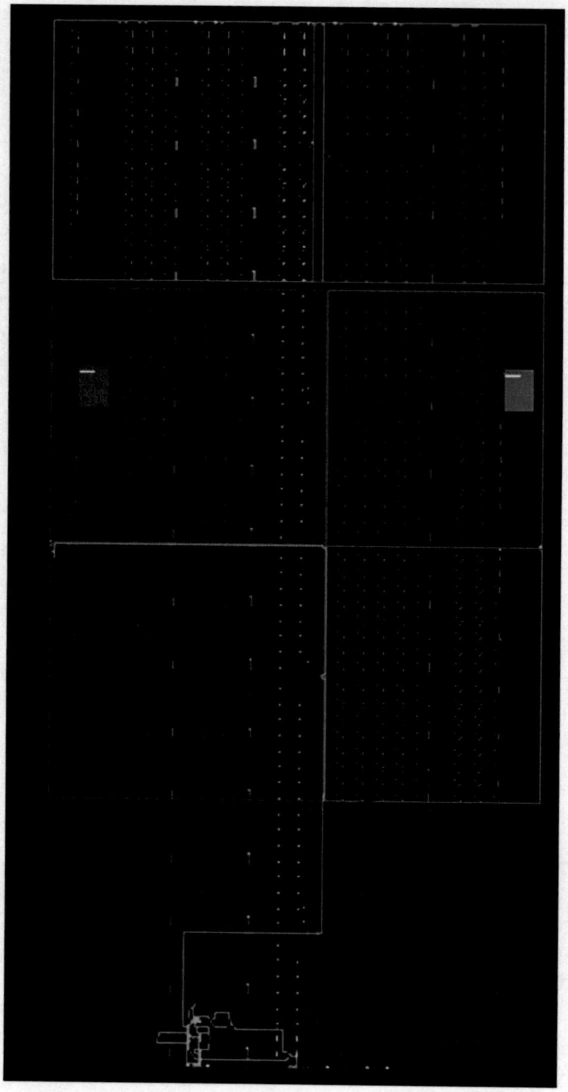

Fig. 18.16 Place and route snapshot for FPGA

18.7 Chapter Summary

Following are few of the important points to conclude the chapter.

1. FPGA is field programmable gate array and called as programmable ASIC.
2. The FPGA important block is configurable logic block which consists of the LUTs and slice registers, and the architecture of FPGA is vendor-specific.
3. During the logic design, the better architecture of the design and partitioning can result into the better performance for the design.

4. The FPGA synthesis differs from the ASIC synthesis. The ASIC uses the standard cells and macros, whereas FPGA uses the LUTs and slice registers with the dedicated blocks and IOs.
5. At the fabric level, the following important steps are used to implement the design
   (a) Logic functionality mapping
   (b) Place and rout
   (c) SDF-based verification
   (d) Signoff STA.

Chapter 19
Prototyping Design

Already we have discussed about the ASIC designs and the design flow and as the design is ready after the logical synthesis and initial floor planning, the prototyping phase can kick off. The better way to prototype design is using the multiple FPGA partitioning and by using few RTL tweaks to have the ASIC to FPGA conversions. The following are few of the points which we need to think during prototype

1. Strategies for the overall prototype and the best FPGA architecture
2. Prototype and system testing plans
3. STA and the interconnect delays incurred during actual testing.

Effectively we try to mimic the ASIC functionality into the actual design and will try to use the FPGA EDA tools to check and test for the desired FPGA functionality. Using the clock speed of the ASIC may not be feasible during the prototype, and the FPGA-based system can be carried out at the lower clock frequency to check for where design fails. If design meets the functionality and constraints for the multiple FPGA prototypes, it almost indicates the design functionality of the ASIC is correct and the design can go through the manufacturing phase.

The following sections discuss about all this in detail and useful to understand the prototyping strategies, issues, and the concepts.

19.1 FPGAs for Prototyping

Consider the design of the 32-bit processor which we have discussed in Chap. 17. Let us try to refer the micro-architecture of the design which is shown in Fig. 19.1.

To check for the functional correctness of the design, ASIC goes through protoyping phase.

V. Taraate, *ASIC Design and Synthesis*,
https://doi.org/10.1007/978-981-33-4642-0_19

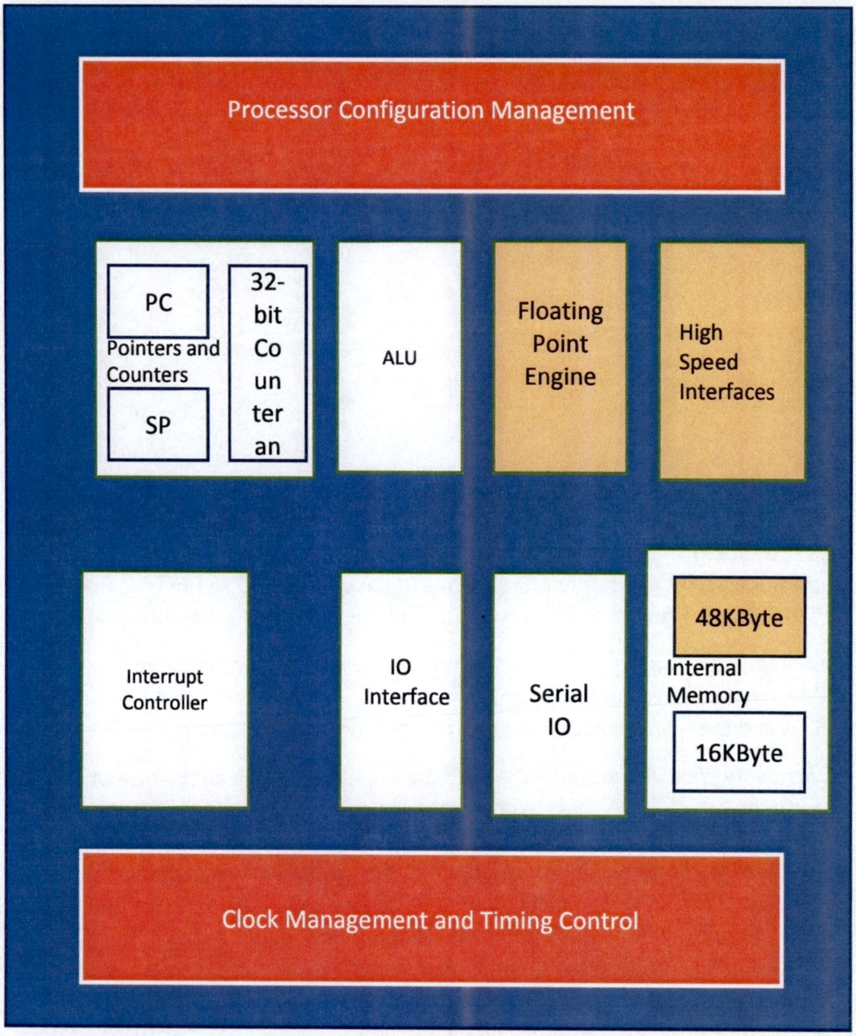

Fig. 19.1 Micro-architecture design

Let us try to identify the goals with reference to following points

1. Prototyping plan
2. Design complexity and the constraints associated for the design
3. The external interfaces and IO requirements and pin multiplexing strategies
4. The suitable FPGA architecture with the IO interfaces
5. Number of FPGAs required to prototype the design
6. The suitable functional and timing proven IPs for the design
7. Documentation on the ASIC to FPGA conversions
8. System testing plan and the documentation.

If the RTL description is designed from the scratch for the FPGA-based designs, then the ASIC to FPGA conversion is not required as the synthesis is performed to have the use of the FPGA functional blocks such as CLBs, IOBs, and BRAM.

Now, try to understand the ASIC prototyping, as the logic design and synthesis, pre-layout STA phase is over the prototype team has information about the overall logic density and the IO requirements. Although the constraints associated with the FPGA interfaces are different as compared to ASICs, the prototype strategies can be finalized to port the ASIC RTL using the multiple FPGAs.

The following section discusses about few of the strategies and their use during the prototyping.

19.2 Synthesis Strategies During Prototyping

As ASICs are faster than the FPGA and logic density is larger, the design partitioning for the million gate SOC is the most important task. The design can be partitioned before synthesis or after synthesis. The prototype team needs to choose the correct approach for partitioning the design.

The truth is design may not run at the SOC speed and it is essential to modify the SOC design into FPGA equivalent resources. So, during the synthesis, it is essential to have clarity about the architecture or initial floor plan, constraints and FPGA resources. Prototyping flow should achieve the better performance as compared to SOC emulation, and for that, the major milestone is synthesis. There are multiple ways in which the synthesis can be performed to achieve the better results. The following are few of the approaches used during the synthesis.

19.2.1 Fast Synthesis for Initial Resource Estimation

If we chose the fast synthesis, then it can be useful for understanding the initial or rough device utilization and the performance at the initial stage. But in such type of synthesis, the full optimization is ignored by the synthesis tool. The reason being the runtime is almost around two or three times. But this can be useful to save the weeks/days time for the complex designs and for the initial design partitioning.

19.2.2 Incremental Synthesis

The incremental synthesis is the better approach for the complex SOC designs. The incremental efforts of P and R tool can be used efficiently while synthesizing the larger density designs. The SOC design sub-blocks or trees can be synthesized separately according to the version changes.

For example, consider the SOC design having 100 sub-blocks and the RTL changes are incorporated in only ten sub-blocks; then during increment synthesis, the tool can synthesize the RTL for the ten sub-blocks. This reduces overall efforts and time during the synthesis phase.

That is if the sub-block or tree architecture is not changed, then synthesis tool ignores this and preserves the previous version. This reduces the weeks/days time for the complex SOC synthesis.

The beauty of the EDA tool like Synopsys Certify [1] or the XILINX P and R tool [2] is that they preserve the hierarchy, previous version logic, placement, constraints, mapping as it is during the re-synthesis if the RTL is not modified. It reduces the turnaround time.

If small portion of the design is modified, then due to incremental synthesis, the design runtime reduces, and P and R tool can use the synthesis results.

The prototype team should be able to use the features of the synthesis and P and R tool. The combined use of these features can reduce the significant amount of time during prototyping. Most important point is that the P and R runtime is always larger than the synthesis runtime for the complex designs. So, the strategy should be use the synthesis and P and R tools in the incremental flow (Fig. 19.2).

The Xilinx EDA tool backend flow is shown in Fig. 19.3.

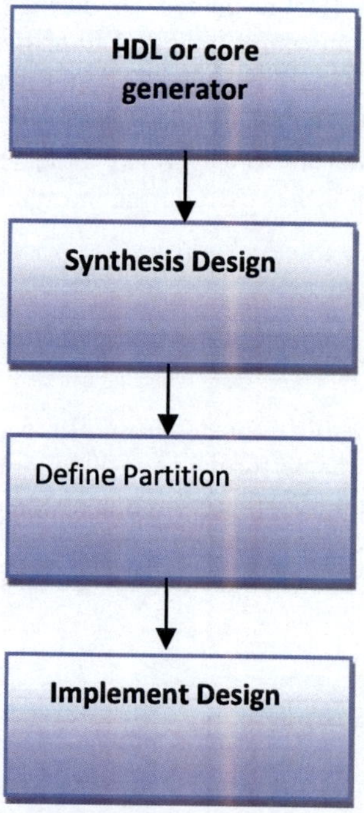

Fig. 19.2 Design synthesis and implementation

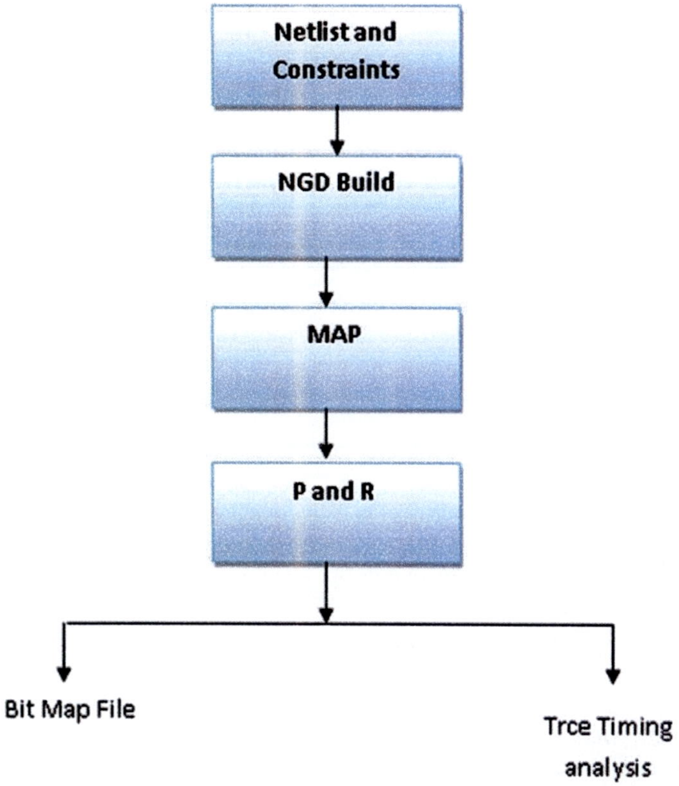

Fig. 19.3 Xilinx backend tool flow

19.3 Constraints During FPGA Synthesis

The Section discusses about the use of tool commands during the FPGA synthesis. The FPGA synthesis commands are listed in Table 19.1.

The following are important steps to be performed during the synthesis using Synopsys DC FPGA

1. Read the Verilog design file.
2. Set the design constraints.
3. Insert the pads.
4. Perform the design synthesis.
5. Execute the replace_fpga command.
6. Write the database.

The sample script for the FPGA synthesis of top_processor_core is shown below

Table 19.1 Commands used during FPGA synthesis

Command	Description
set_port_is_pad <port_list> <design_list>	The command is useful to place attributes on the list of ports specified in command. Attribute allows dc to map IO pads
set_pad_type <type of pad> <port_list>	The command is used to choose the type of the pads to which design is to be mapped
insert_pad	The command is used to insert the pads
replace_fpga	The command is used to convert the synthesizable FPGA database to the schematic. Instead of visualizing the schematic having CLBs, IOBs, the schematic consists of gate

```
------------------------------------------------------------------------------
dc_shell > read –format verilog top_processor_core.v
dc_shell> create_clock clock –name clk –period 10
dc_shell> set_input_delay 2 –max – < list all the input ports using the same
command and required delay attribute>
dc_shell > set_port_is_pad
dc_shell> insert_pad
dc_shell> compile –map_effort high
dc_shell> report_timing
dc_shell> report_area
dc_shell> report_cell

------------------------------------------------------------------------------
```

The timing report consists of the timing path information and the data required time, data arrival time, slack.

Area report gives the list of following:

```
------------------------------------------------------------------------------
Number of ports
Number of cells
Number of nets
Number of references
Combinational area
Non-combinational area
Net Interconnect area
Total cell area
Total area
------------------------------------------------------------------------------
```

To get the information about the FPGA resources, the following command can be used

dc_shell > report_fpga–one_level

It gives the following information about the use of FPGA resources

---

Function Generators:

Number of CLB
Number of ports
Number of clock pads
Number of IOB
Number of flip flops
Number of tri state buffers
Total number of cells

To write the netlist in the database format, use the command

dc_shell > write-format db–hierarchy–output top_procesor_core.db

The synthesizable database (netlist) and timing information can be used by the place and route tool.

19.4 Important Considerations and Tweaks

The following are few of the important considerations useful during the prototyping to get the FPGA equivalent logic.

1. **Gated Clock instantiation**: The gated clock structure for the SOC may not be matched with the FPGA equivalent structure, and hence, it is essential to modify the RTL to infer the gated clock structure (Figs. 19.4 and 19.5).

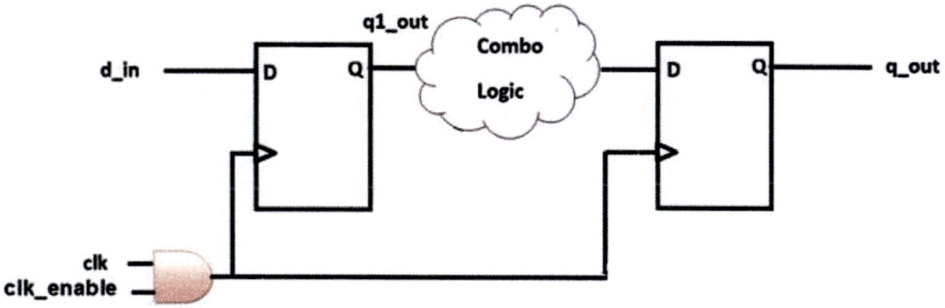

Fig. 19.4 Gated clock used for the design

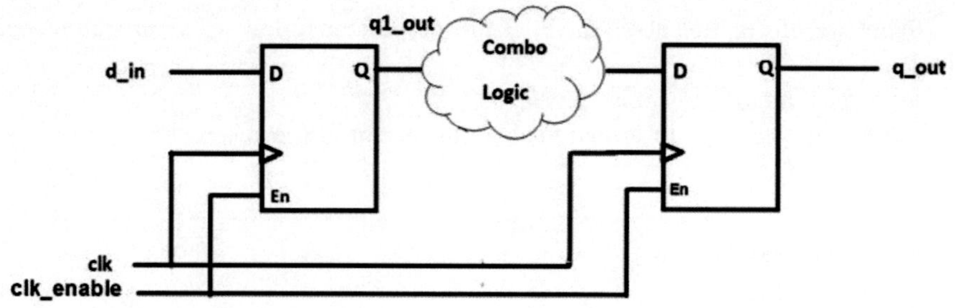

Fig. 19.5 FPGA equivalent clock gating

2. **SOC IPs**: Most of the time IPs with the RTL design is not available and hence it is essential to have the FPGA equivalent of such IPs.
3. **ASIC/SOC memories**: The memory structure for the ASIC or SOC is not identical with the FPGA memories, and hence, it requires the modification during the prototype stage.
4. **Top-level pads**: As FPGA tool doesn't understand about the instantiation of the pad, hence it is essential to modify them during the prototype. As it does not handle the IO PAD in the RTL and infers the FPGA PAD, so it needs to leave the pads out with dangling connections inactive or to the top-level boundary. For the prototype, replace each IO pad instance with synthesizable model of FPGA equivalent.
   The model should have the logical connections at the RTL level and that can be done by writing small piece of code in the RTL. For the efficient prototype, prepare the SOC pad library. The basic FPGA IO cell is as shown in Fig. 19.6.
5. **IPs in the netlist forms**: The netlist form may not be the FPGA equivalent and hence needs modification during prototype.
6. **Leaf cells**: Leaf cells from the ASIC library may not be understood by the FPGA, and hence, it needs modifications.
7. **Test circuitry**: The built-in self test (BIST) and other test or debug circuit need to have the FPGA equivalent and hence need the modification.
8. **Unused inputs**: For the unused input pins, it is essential to tweak the RTL.

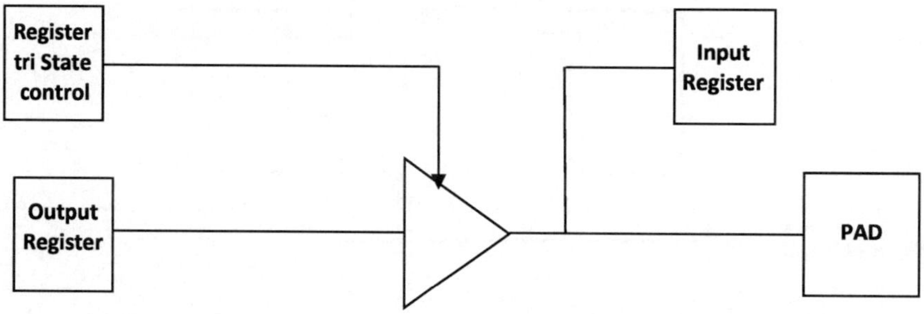

Fig. 19.6 FPGA basic IO cell

9. **Generated clocks**: During prototype to achieve the better performance, the generated clocks need to be modified by its FPGA equivalent.

19.5 IO Pad Synthesis for FPGA

As FPGA tools do not understand about the instantiation of the pads, hence it is essential to modify them during the prototype. As it does not handle the IO pad in the RTL and infers the FPGA pad, so it needs to leave the pads out with dangling connections inactive or to the top-level boundary. For the prototype, replace each IO pad instance with synthesizable model of FPGA equivalent.

The model should have the logical connections at the RTL level and that can be done by writing small piece of code in the RTL. For the efficient prototype, prepare the SOC pad library. The basic IO cell for the FPGA is shown in Fig. 19.6.

Use the following commands using Synopsys DC. For more information on FPGA synthesis, refer Sect. 18.2 (Fig. 19.7).

```
dc_shell > set_port_is_pad
dc_shell> insert_pad
dc_shell> compile –map_effort high
```

19.6 Prototyping Tools

The ASIC prototyping is achieved by using industry standard leading tools like Design Compiler FPGA. The EDA tool is used to have ASIC prototyping for high denisty FPGA designs. The design compiler is industries leading EDA tool and used to get best optimal synthesis result and best timing for the FPGA synthesis. The basic flow for the ASIC prototyping is shown in the (Fig. 19.7)., and in the subsequent chapters, we will discuss about the ASIC prototyping using multiple FPGAs and how we can achieve the efficient ASIC prototype (Fig. 19.7).

19.7 Chapter Summary

Following are few of the important points to conclude the chapter

1. Multiple FPGAs are used to prototype the design.
2. The incremental synthesis is the better approach for the complex SOC designs.
3. The ASIC prototyping is achieved by using industry standard leading tools like Design Compiler FPGA.

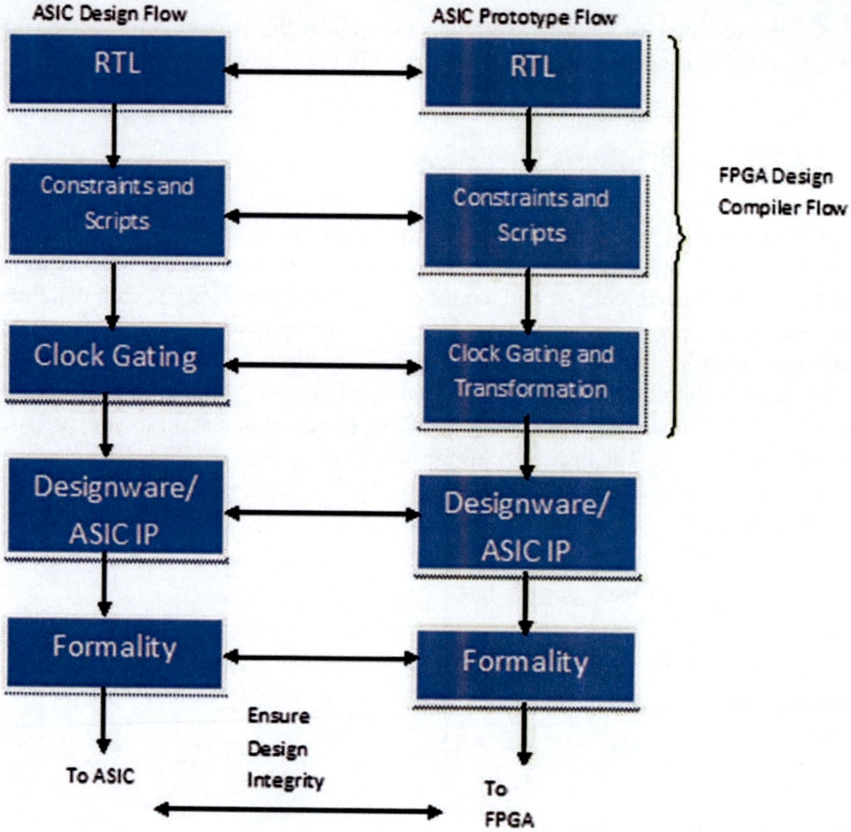

Fig. 19.7 ASIC prototype flow

4. The gated clock structure for the ASIC may not be matched with the FPGA equivalent structure, and hence, it is essential to modify the RTL to infer the gated clock structure.
5. As FPGA tools do not understand about the instantiation of the pads, hence it is essential to modify them during the prototype.

Chapter 20
Case Study: IP Design and Development

As the ASIC design functionality is complex, the design need to have the use of various IPs. The important considerations while design of IPs or use of IPs is based on the IO requirements and functional and timing requirements for the design. In such circumstances, it is helpful to the design team to have the understanding of the various kinds of available IPs. The next subsequent section discusses the IP design and development strategies and their reuse during the design.

20.1 IP Design and Development

Consider the design of the SOC which consists of the various functional blocks such as the processors, floating point engine, H.264 encoder, DDR controller, etc. For the quick turnaround of the design general industry practice is to have the use of IPs. Consider that the DDR memory controller IP is available in the market then instead of design of the memory controller from scratch, the industry practices are used for the available functional and timing-proven IPs during the design.

IPs are available in the following format, and the prototype team needs to use the IPs during the design cycle at the different stages.

1. RTL Source code of IP: Open source code or the license version of the IP source code is available. The source code using VHDL or Verilog is available.
2. **Soft IP**: This type of IP cores is sometime encrypted versions, and they need to have some processing during the design and reuse.
3. **IPs in the netlist form**: They are available in the form of the pre-synthesized netlist of the SOC components or Synopsys GTECH.
4. Physical IP: They are also called as hard IPs, and they are pre-laid-out by the foundry.

Most of the time during ASIC design we need to use functional and timing proven IPs!

V. Taraate, *ASIC Design and Synthesis*,
https://doi.org/10.1007/978-981-33-4642-0_20

5. Encrypted Source Code: The RTL is protected with the encrypted key and must be decrypted to get the RTL source.

20.2 What We Consider During the IP Selection

The following are important points which we consider during the selection of IPs

1. Functional requirements and the features supported by the available IPs.
2. The IOs and other high-speed interfaces for the IP.
3. Format in which IPs are available. That is whether the IP tweaking to boost the performance is possible or not?
4. What kind of the configuration environment IP is having.
5. What are the debug and test features available in the IP.
6. What kind of documentation is provided by the IP vendor?
7. What are the electrical characteristics for which IP can be used?
8. What is the environment in which IP can be used?
9. Different clocks and power domain for the IPs.
10. What are the timing characteristics and IO delays for the IPs?

By considering this, we will try to select the IP.

20.3 Strategies Useful During the IP Design

Following are few of the strategies which can be useful during the design of IP. Although IP design and verification is time consuming phase if the design demands the new functional implementations, then it is mandatory to have the IP design and development. For example, new standard is available in the market and in such scenarios the design houses may work on the IP design and development.

1. **IP Design and reuse**

Most of the SOC design team always uses the third-party functional and timing-proven IPs. During the design of the complex ASICs for quick turnaround, the IPs can be reused. Use of hard or soft IPs can be used during the design and prototype phase and the reuse is helpful to achieve.

1. Focus on the design of additional supported features for quick turnaround.
2. Reduces the time to market.
3. Design team will be able to spend more time to have low-power and high-speed designs.
4. Design team will be able to play around using the multiple clock domain and multiple power domain designs.

5. The physical design challenges such as fixing the timing violations needs more time during the physical design. So if IPs are used, this time is reduced significantly.
2. **Hardware–Software co-design**

This is also called as design partitioning; the design has to be partitioned into hardware and software. The important point of consideration is while partitioning the design; how parallel execution needs to be incorporated in the design? In the present scenario as SOCs are complex, the functionality can be implemented using the parallelism in the design which in turn can improve the design performance. The complex computational task or algorithms need to be partitioned during the design partitioning phase. Most of the complex computational blocks need to be implemented using hardware. Design partitioning is important and decisive phase to define what need to be implemented using software? And what need to be implemented using hardware?

For example, consider the design of video decoders which need multiple frame support. The video decoder can be efficiently implemented using hardware, and even the parallelism can be incorporated for the few decoder features. The high computational DSP functional blocks which need filters like FFT, FIR and IIR or high-speed multipliers can be effectively and efficiently implemented using hardware.

Let us consider the scenario of protocol implementation, most of the protocols like Ethernet, USB, and AHB can be efficiently implemented using hardware software co-design. These algorithms should be functional and timing proven. This can have advantage to overcome and to reduce latency in the design. For most of the protocol implementation, it is essential to consider.

The major challenge in the hardware software design portioning is the analysis of throughput and power requirements. For example, consider the scenario in SOC design where fixed length packets need to be transferred over the fixed time interval. If the design is implemented by using hardware, then care needs to be taken such that there should be minimum interaction between the hardware and software. To minimize the interaction between hardware and software, the strategy can be used by using FIFO buffers and timers.

3. **Interface details and timing requirements**

For every IP, it is essential to have the functional and timing-proven bus interfaces. In most of the applications, Advanced High Speed Bus protocols are used. These protocols need to be validated for the functional and timing correctness of the design. IO interfaces need to be targeted for the high-speed data transfer. There are many different kinds of IO interfaces used in SOC designs. These IOs can be general-purpose, differential IOs, and high-speed IOs.

Reset clock requirements:

Clock distribution network is used to provide the uniform clock skew to all the registers in the SOCs. The clocking policy plays the crucial role in overall design performance. The uniform clock skew can be achieved by using the suitable clock tree by using clock tree synthesis. Use of single clock structure or multiple clock domain structure needs to be decided at the architecture level. Also the uses of synchronous or asynchronous logic need to be defined at the architecture level. Reset can be asynchronous or synchronous and need to be defined at the architecture phase of SOC.

4. ***EDA tool and license requirements***

Choose the required necessary EDA tools and licenses for FPGA prototyping of a SOC and for ASIC porting. The most industry standard tools are

Simulator: Questasim, VCS
Synthesis: Synpilfy pro and Synopsys DC
STA: prime time (Synopsys PT).

5. ***Developing the required prototyping platform:***

For SOC and IP validation, use the necessary prototyping and development platform. Prototyping platform can consist of use of multiple FPGA boards to realize and validate SOCs, IP required, DSP functionality required, memories and general-purpose processors required. The availability of desired prototyping boards with the necessary interfaces to realize SOC and use of debug or testing setup.

Most of the SOCs are tested by using the test setup consisting of available EDA tools and logic analyzers. At the start of the SOC design cycle, architect analyzes the design and functional requirements and according to the requirement of speed and estimation of gate count the prototyping platform can be designed. Here the overall important factors are time to market, budget allocation, and design time requirements. If DSP capabilities are available in FPGA, then it is wise to implement the DSP functionality on FPGAs.

6. ***Developing the test plan***:

For complex gate count SOCs, the necessary test cases need to be developed with the required test vectors. The features can be extracted using top-level functional specifications, and the required test cases can be documented in the test plan document. The test vectors developed can have significant impact on the quality of the verification to achieve the coverage goals. The test cases can be documented as basic, corner, and the random test cases. The constrained random verification with the required coverage goals can be achieved by using the required necessary test cases.

7. ***Developing the verification environment***:

Use the verification languages like Verilog and high-level verification languages like System Verilog or System C; for early detection of bugs and to achieve the coverage goals. The verification planning to improve the overall design quality by capturing the bugs during early design cycle is always crucial in the large gate count SOC designs. The overall objective is to achieve the required and designed functionality in less time. The verification environment needs to be built to achieve the coverage goals. The verification architecture can have the necessary bus functional models and the drivers, monitors and scoreboards for robust checking of the design specifications. The overall verification planning and creation of environment is with goal to achieve the automation to minimize the time requirement to complete the functional checks in the lesser amount of time duration.

20.4 Prototyping Using Multiple FPGA

Consider the SOC design which has the processor for the general-purpose computing, DDR3 memory controller and video encoder and decoder IP. If the design need is of 2, 00,000 logic gates, then the design cannot fit on the single FPGA of Artix-7. In such circumstances, we need to use the design partitioning to target the design using multiple FPGAs. For most of the SOCs, we need to target the prototype using the multiple FPGA architecture. The FPGAs can be connected using ring- or star-type topology. The Fig. 20.1 describes logic on the multiple FPGAs using the cable connectivity (Fig. 20.1).

Following are few of the important recommendations useful during the prototype using multiple FPGAs.

1. **Have better understating of the design**: Try to understand the analog and digital functionality for the design and partition the design into the analog and digital design domains. Use the partitioning tools to have the better results. The automatic partitioning tools can be used to have the better partitioning of the design across sequential boundaries.
2. **Analog functionality and the additional interfaces**: As FPGA is good candidate to realize the digital design, but practically the design has both analog and digital

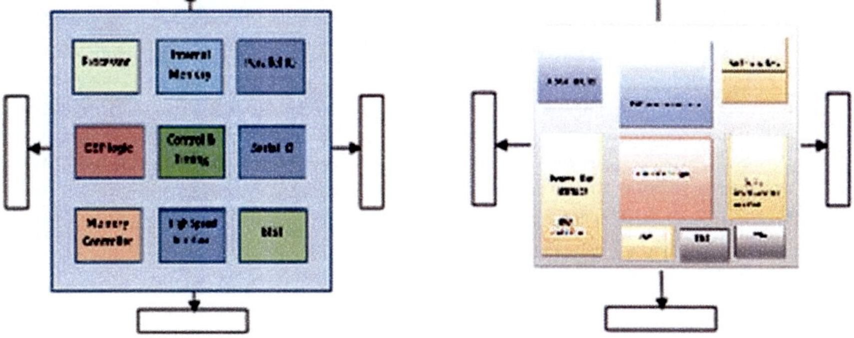

Fig. 20.1 Multiple FPGA during SOC prototype

blocks. So try to choose the additional daughter boards to interface the ADC and DAC.

3. **Efficient use of the resources**: Try to have strategy while performing the partitioning to allow the EDA tool to have maximum 70% of the FPGA resources. This will allow the prototype team to add the BIST and debug logic during the board bring-up phase.
4. **Requirement of IOs and pin multiplexing**: The speed of IO is an important factor which decides the overall performance of the prototype. There are additional multiplexing strategies need to be deployed for the multiple FPGA designs.
5. **Clocking strategies**: Depending on the requirement of the star, ring topology, it is essential to think about the clocking strategies for the multiple FPGA designs. The clock skew and other board delays need to be thought during the debug and the test phase.
6. **IO interfaces**: At the SOC architecture level, the decision should be made about the prototype features requirement. Always it is better choice to consider about the IO speed, IO voltage, bandwidth, clock and reset network, external interfaces while designing the prototype using the single or multiple FPGAs!
7. **FPGA connectivity**: The prototype team needs to think about the ring-, star-, or mix-type connectivity for the prototype using the multiple FPGA. Following are few of the highlights

   (a) **Ring-type connectivity between FPGAs**

In such type of arrangement, the multiple FPGAs are connected to form the ring.

In such type of connectivity, it increases the overall path delay. As the signal is passing through the FPGA, the equivalent prototype logic can resemble to priority logic. This type of the connectivity has slower speed as compared to other type of boards.

If we try to visualize the ring-type connections, then at high level we can think about the pin connection using such type of inter FPGA connectivity. The wastage of IOs cannot be limited in such kind of the connectivity. The FPGAs are at the down side; IOs will be wasted, and it is additional overhead to the board designer and board layout team to connect these IOs to high impedance states.

(b) **Star connectivity**:

This type of inter FPGA connectivity is faster as compared to the ring arrangement due to the direct connections with the other FPGA. For the better prototype performance, use the high-speed interconnects between the FPGAs and configure the unused pins as high impendence state.

c. **Mixed connectivity**:

During the board design and layout, we may use the mix of the ring-type connections and star connectivity. Such type of connectivity can have the moderate performance.

The boards available in the market from vendors have fixed connectivity and may not be suitable during prototyping as they don't match the specifications and requirements. Under such circumstances depending on the design complexity, it is better to choose interface connectivity for better prototype performance.

20.5 H.264. Encoder IP Design and Development

Let us try to have strategies to develop the IP for the H.264 encoder design. The architecture is shown in Fig. 20.2: H.264 Encoder Architecture.

The following we need to think during the micro-architecture design for the H.264 encoder.

20.5.1 Features and Micro-architecture Design Strategies

1. **Video formats supported**: Various video formats supported by the encoder that is SD size, HD size, etc.

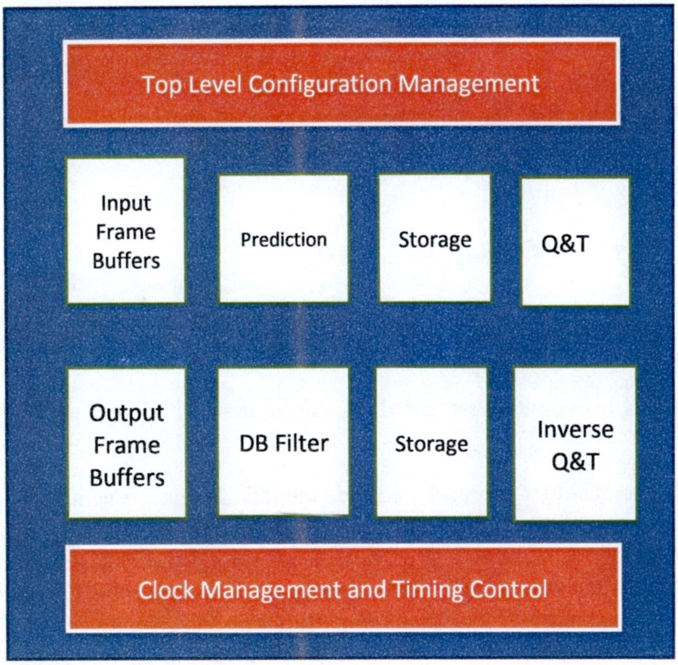

Fig. 20.2 H.264 encoder architecture

2. **Maximum frame size**: What is the maximum frame size for example 1920 × 1080 predictive frame?
3. **Timing requirements**: The number of framers per second supported for the processing and the overall clocking requirements.
4. **Processing of video data**: Strategies while processing of the video data that is hardware and software co-design, configuration management.
5. **Design partitioning**: To have the better understanding have the strategies for the manual partitioning of the design at.
   (a) Functional level
   (b) Interface level
   (c) Across multiple clock levels
   (d) Across multiple power domains
6. **Functional understanding and interfaces**: For better design outcome have the understanding of the
   (a) Block-level functionality of prediction (intra, inter), quantization, transform, inverse quantization, inverse transform and deblocking filter blocks.
   (b) Sketch the micro-architecture representation for each functional block
   (c) Document the interfaces for each block
7. **External interfaces**: Have the better strategy to have the external high-speed interfaces and interfaces for the communication.
8. **Memory requirements**: Have the better strategies to have the memory requirements and buffering mechanism for the design.
9. **EDA tools required:** Have the documentation on the budget requirement for the EDA tools and debug test setup tools.

20.5.2 *Strategies During RTL Design and Verification*

Use the following strategies during the RTL design and verification of the H.264 encoder

1. **RTL Design**: Following we can think during the RTL design phase of H.264 encoder.
   (a) Use the modular approach and code the RTL for each functional block. Try to eliminate the combinational logic hierarchy for the better optimization.
   (b) Have the test synthesis strategies placed during the RTL design for each module that will give the DFT-friendly architecture for the H.264 encoder.
   (c) Have the multiple clock domains partitioning and groping and then specify the hierarchy for the design. Code the RTL for the various clock domains using separate module and then have strategies to deploy the synchronizers in the data and control paths.

(d) If low-power-aware architecture is the requirement, then try to use the low-power strategies during the RTL design level.
(e) Top-level design and integration should use the readable naming conventions and should have the interface at the sequential boundaries for the better timing, and this care should be taken by every RTL design team member.

2. **RTL Verification**: Following we can think during the RTL verification phase of H.264 encoder.

   (a) Have the verification plan in place for the block- and top-level verification
   (b) Have the automatic testbenches at the module level and top level; with the goals to have the better coverage
   (c) Try to have the test cases and test vectors for the block-level functionality and top-level functionality
   (d) Have the top-level verification strategies and architecture to find the issues in the RTL design and to notify the RTL team.

20.5.3 Strategies During Synthesis and DFT

Use the following strategies during the logic synthesis of the H.264 encoder

1. **Bottom-up compilation**: Use the bottom-up compilation strategies.
2. **Block-level synthesis**: Perform the synthesis for each functional block using the block-level constraints.
3. **Top-level synthesis**: Perform the top-level synthesis using top-level constraints.
4. **Performance improvement**: Improve the area and speed performance for the design using the various techniques discussed in Chaps. 11–13.
5. **DFT and scan insertion**: Have the strategy to use the full scan or partial scan to detect the faults in the design.

   (a) Use the DFT techniques to find the stuck at fault and fault coverage.
   (b) Try to find the test violation and report to the team.

20.5.4 Strategies During Pre-layout STA

Use the following strategies during pre-layout STA.

1. Use the STA tool with goal to report all violations in the design.
2. Fix the setup violations in the designs using techniques specified in Chap. 15.
3. Use the various performance improvement techniques to meet the timing goals.

20.5.5 Strategies During Physical Design

Use the following strategies during physical design.

1. Use the floor planner with the floor plan manager to exchange the information between the front-end and back-end tools.
2. Have the placement and routing with the goal to optimize the design to yield the clean timing.
3. Fix up the timing violations setup and hold for the timing paths where the violations are reported.
4. Use the LVS and DRC checks to report the violations and have strategies to fix them.

20.6 ULSI and ASIC Design

If we consider the toddy's world, the design demands the lot of intelligence in the chips. The ASICs which may be used in the aerospace, communication, video processing and general processing applications may need to have the self and debug test mechanisms and should have the intelligence to using the AI-based designs.

As the technology is advancing and the shrinkage is demand, there is evolution of the processes used during the manufacturing of the ASICs. If we observe carefully the growing consumer market the demand of use of AI and ML-based designs to embed the intelligence in the chips.

The Very Large Scale Integration which we treat as the design with the billions of the logic gates and such kind of ASICs are manufactured by foundries to have better reliability, durability. The advances of the technology in the past one decade where the process node has shrunk enough below 10 nm have imposed lot of challenges in the overall design and manufacturing processes. Few of these are due to the

1. Low voltage levels.
2. Noise impact.
3. Interconnect delays.
4. Foundry laid rules.

So to cope up with the requirements, the design and manufacturing processes have the evolution and the ULSI-based designs is the need of the market. ULSI is Ultra Large Scale Integration and used to have the density of the few billion logic gates which operate at the lower voltage levels might be in the range of 0.8 Volts to the 1.5 Volts.

Due to need of the lower core and logic power, the design has various issues during the initial architecture finalization to achieve the timing and performance.

For this kind of the ASIC, the noise is one of the biggest bottlenecks and the major issues during the place and route while meeting the design and optimization constraints. The actual wire delays and the optimization of the design for the performance is one of the challenging tasks for the billion gate ASICs.

Even the internal data integrity and synchronization of the multiple clock domain complex modules is the major challenge in the design and to have the clean layout.

With the shrinking process node, the evolving algorithms and processes and the EDA-based environment availability is another challenge as for the ULSI-based designs the NRE cost for the design, and test is very high and even the EDA industry is going through the evolution of the algorithms which are useful to have the ASIC-based designs at 2 or 3 nm.

This demands the intelligence which needs to be embedded in the design, and many chip companies are working in the area of AI- and ML-based designs.

20.7 Chapter Summary

Following are important points to conclude the chapter.

1. During the design of the complex ASICs for quick turnaround, the IPs can be reused.
2. Functional and timing-proven IPs are used during the design and prototype.

Appendix A

Verilog is case-sensitive, and the important Verilog 2005 constructs are listed below:

V. Taraate, *ASIC Design and Synthesis*,
https://doi.org/10.1007/978-981-33-4642-0

1. module declaration

```
/////////////////////////////////////////////////////////////////////////////////
module comb_design ( input wire a_in, b_in, output wire
y1_out,y2_out, output reg [7:0] y3_out );

//Concurrent and sequential statements and assignments

endmodule
/////////////////////////////////////////////////////////////////////////////////
```

2. Continous assignment (neither blocking nor non-blocking)

```
assign y1_out = a_in ^ b_in;// net type is wire
```

3. ***always@**** // Combinational procedural blocks

```
/////////////////////////////////////////////////////////////////////////////////

always @*
begin
// blocking assignments or sequential constructs and net type reg
end
/////////////////////////////////////////////////////////////////////////////////
```

4. ***always@ (posedge clk)*** // sequential procedural block sensitive to positive edge of clock

```
/////////////////////////////////////////////////////////////////////////////////
```

***always @(posedge** clk)*
begin
//synchronous reset and assignments
//non-blocking assignments or sequential constructs (net type reg)
end
//

5. ***always@ (posedge clk or negedge reset_n)*** // sequential procedural block sensitive to positive edge of clock

//

***always @(posedge** clk or **negedge** reset_n)*
begin
//asynchronous reset and assignments
*//non-blocking assignments or sequential constructs (net type **reg)***
end
//

6. ***always@ (negedge clk)*** // sequential procedural block sensitive to negative edge of the clock

//

***always @(negedge** clk)*
begin
*// non-blocking assignments or sequential constructs and net type **reg***
end
//

7. Multiple blocking (=) assignments in the procedural block

//

```
begin
    tmp_1 = data_in;
    tmp_2 = tmp_1;
    tmp_3 = tmp_2;
    q_out = tmp_3;
end
```

//

8. Multiple non-blocking (<=) assignments in the procedural block

//

```
begin
    tmp_1 <= data_in;
    tmp_2 <= tmp_1;
    tmp_3 <= tmp_2;
    q_out <= tmp_3;
end
```

//

9. Sequential construct ***if –else*** within always procedural block

//

```
if(condition)
//assignment or expression
  else
//assignment or expression
end
```

//

10. Sequential construct ***case--endcase*** within always procedural block

```
////////////////////////////////////////////////////////////////////////////////

  case (sel_in)

// conditions and expressions

  endcase
////////////////////////////////////////////////////////////////////////////////
```

11. Sequential construct ***casex--endcase*** within always procedural block

```
////////////////////////////////////////////////////////////////////////////////


      casex (sel_in)

              // conditions and expressions

      endcase
////////////////////////////////////////////////////////////////////////////////
```

12. Sequential construct ***casez--endcase*** within always procedural block

```
////////////////////////////////////////////////////////////////////////////////
     casez (sel_in)

      casex (sel_in)

              // conditions and expressions

      endcase
////////////////////////////////////////////////////////////////////////////////
```

13. Procedural block ***initial***

```
////////////////////////////////////////////////////////////////////////////////
```

```
initial
begin
//assignments with non-synthesizable intent
end

////////////////////////////////////////////////////////////////////////////////////
```

For the other constructs, please refer Verilog 2005 language reference manual!

Appendix B

The important Synopsys Design Compiler (DC) commands useful during synthesis are listed in the table below

DC command	Constraint type	Description of command
set_max_transition	DRC	Used to define the largest transition time
set_max_fanout	DRC	Used to set the largest fanout for the design
set_max_capacitance	DRC	Used to set the maximum capacitance allowed for the design
set_min_capacitance	DRC	Used to set the minimum capacitance allowed for the design
set_operating_conditions	Optimization constraints	Used to set the PVT conditions as it affects on timing
set_load	Optimization constraints	Used to model load on output port
set_clock_uncertainty	Optimization constraints	Used to define the estimated network skew
set_clock_latency	Optimization constraints	Used to define the estimated source and network delays
set_clock_transition	Optimization constraints	Used to define the estimated input skew
set_max_dynamic_power	Power constraints	Used to set the maximum dynamic power
set_max_leakage_power	Power constraints	Used to set the maximum leakage power
set_max_total_power	Power constraints	Used to set the maximum total power
set_dont_touch	Optimization constraints	It is used to prevent the optimization of mapped gates

V. Taraate, *ASIC Design and Synthesis*,
https://doi.org/10.1007/978-981-33-4642-0

Bibliography

https://www.physicaldesign4u.com/2019/12/floorplanning-floor-planning-is-art-of.html

http://vlsibyjim.blogspot.com/2015/03/power-planning.html

https://anysilicon.com/ic-layout-an-overview/

www.synopsys.com. Power compiler reference manual. Synopsys Inc

www.ieee.org. IEEE1801 low power design standard

www.synopsys.com. Guidelines and practices for successful logic synthesis version 1998.08, Aug 1998

www.synopsys.com. Synopsys timing constraints and optimization user guide, version D-2010.03

www.springer.com. Digital logic design using verilog (https://www.springer.com/in/book/9788132227892)

V. Taraate, *ASIC Design and Synthesis*,
https://doi.org/10.1007/978-981-33-4642-0